A Practical Guide to Alterations and Extensions

WAWK

A Practical Guide to Alterations and Extensions

Andrew R. Williams

E & FN SPON
An Imprint of Chapman & Hall

London · Glasgow · Weinheim · New York · Tokyo · Melbourne · Madras

**Published by E & FN Spon, an imprint of Chapman & Hall,
2–6 Boundary Row, London SE1 8HN, UK**

Chapman & Hall, 2–6 Boundary Row, London SE1 8HN, UK

Blackie Academic & Professional, Wester Cleddens Road, Bishopbriggs, Glasgow G64 2NZ, UK

Chapman & Hall GmbH, Pappelallee 3, 69469 Weinheim, Germany

Chapman & Hall USA, 115 Fifth Avenue, New York, NY 10003, USA

Chapman & Hall Japan, ITP-Japan, Kyowa Building, 3F, 2–2–1 Hirakawacho, Chiyoda-ku, Tokyo 102, Japan

Chapman & Hall Australia, 102 Dodds Street, South Melbourne, Victoria 3205, Australia

Chapman & Hall India, R. Seshadri, 32 Second Main Road, CIT East, Madras 600 035, India

First edition 1995

© 1995 Andrew R. Williams

Typeset in 10/12 pt Plantin by Cambrian Typesetters, Frimley, Surrey
Printed in Great Britain by The Alden Press, Osney Mead, Oxford

ISBN 0–419–20080–0

A catalogue record for this book is available from the British Library

Library of Congress Catalog Card Number: 95–67605

∞ Printed on permanent acid-free text paper, manufactured in accordance with ANSI/NISO Z39.48–1992 and ANSI/NISO Z39.48–1984 (Permanence of Paper).   [Paper = Magnum, 70 gsm]*

For
Geraldine, my wife
and
Thomas Grenville Williams, my late father

About the author

Andrew R. Williams is a Principal Surveyor in private practice. The practice was first established in 1977 and since then has provided a wide variety of quantity surveying and building surveying services.

By the same author:

Domestic Building Surveys, E & FN Spon, London. A Practical Guide to Single Storey House Extensions, Building Trades Magazine.

Contents

Preface   ix
Acknowledgements   xi

Part One: Introduction
1   A general guide to drawing the plan   3
2   The need for control   9
3   The Local Authority   11

Part Two: More on Planning
4   Council planning policies   15
5   Is planning approval necessary?   23
6   Applying for planning permission   31
7   Conservation areas/listed buildings/tree preservation orders   39
8   Other design aspects for consideration   49

Part Three: More on Building Control
9   More on building control   67
10 Is building control approval required?   71
11 Making the building control submission   73

Part Four: Building Construction
12 Notes concerning new Approved Document L – Conservation
    of Fuel and Power (1995 Edition)   85
13 The foundations   91
14 External walls   103
15 Internal walls   117
16 Ground floors   121
17 Timber upper floors   127
18 Flat roofs   129
19 Pitched roofs   137
20 Finishes etc.   151
21 Ventilation to habitable rooms, kitchens, bathrooms and WCs   153
22 Staircases   155
23 Drainage and plumbing   159
24 Other building control aspects   163

Part Five: Mainly for Consultants/Conclusion
25   Preparing to meet the client   169
26   Meeting the client   171
27   Conclusion   179

Appendices
Appendix A: Checklist of items not addressed or often missed
             off plans   183
Appendix B: A standard specification   185
Appendix C: Building construction terminology   195
Appendix D: General notes and standard conditions of
             engagement   201

Index   206

Preface

In the UK, home ownership has traditionally been the ambition of a very large section of the population. The concept of the property-owning democracy is something that has been fostered by a large number of British politicians for years. Although tax concessions on housing are now being slowly eroded, there is no doubting that home improvements will continue to provide work for a large number of tradesmen builders, home improvement companies and professionals.

The basic skeleton of this book was created many years ago when I prepared an 'in-house' manual containing my standard specifications, standard details and standard letters for use in my office. Although produced long before the introduction of BS 5750, the reason for creating the manual was the same. I wanted to ensure that all drawings produced in my office were prepared to a uniform standard and that procedures could also be standardized. The first edition of this book was published by *Building Trades Journal* as part of their Practical Guide series and delighted under the snappy title of *A Practical Guide to Single Storey House Extensions*.

Although the basic principles described in the previous publication remain valid, major revisions to the Building Regulations have now made some of the technical information in the original very out of date. In addition, since the first edition was published, I have been pleasantly surprised to receive several telephone calls praising the original but asking if I ever intended to produce a more advanced version of the book. Thus motivated, I have taken the opportunity to expand the contents.

The book is still aimed at the same target audience, namely:

(a) Smaller building companies/tradesman builders.
(b) Draughtspersons/junior architectural technicians.
(c) Surveyors/building consultants.
(d) Students.
(e) Householders considering altering their homes, who want to 'read around' the subject.
(f) The keen DIY enthusiast.

Whilst the tradesman builder might be an expert in their own speciality, in my experience, most do not know a great deal about planning and building control procedures. The book should therefore provide them with a useful guide, especially concerning matters such as Permitted Development and the effects of modern Building Regulations.

Students, technicians and consultants will hopefully find the technical details of assistance when preparing their own plans.

For the householder, this book will hopefully explain the basics of the planning and building control system in England and Wales and give them an insight into simple building construction. Then, if and when they engage a building surveyor to design their extension, they will understand what work is being carried out on their behalf. I have stressed basics, because the book is only intended to cover simple domestic situations.

For the DIYer, a challenge is always a challenge and I have no doubt that this book will provide enough information for the avid DIYer to prepare his own plans for the simpler type of home extension and submit them to their Local Authority for planning and building control approvals. Naturally enough, a little knowledge can be a dangerous thing, and I would not envisage that a DIYer would attempt to design a major building.

Andrew R. Williams FRICS, FCIOB, FBEng, FIAS, MIBC
Andrew R. Williams & Co.
Chartered Quantity Surveyors
Corporate Building Engineers
First Floor, HSL Buildings
437 Warrington Road
Rainhill, Merseyside, L35 4LL

Acknowledgements

The author wishes to thank David Tierney (Principal Building Control Manager of Halton Borough Council), Ian Davis (NHBC Director of Technical Standards), Steve Le Guen and Leonard Fernley for their help and assistance whilst preparing the manuscript for this book.

As many of the technical illustrations were derived from manufacturers' technical catalogues, the author also wishes to thank Ibstock Building Products Ltd., Owens Corning (formerly Pilkington Insulations), Willan Building Services Ltd., Marley Building Materials Ltd., Alfred McAlpine Minerals (Penrhyn Slate Quarries) and WL Computer Services. Last but not least, the author wishes to thank Halton Borough Council and St. Helens Borough Council for permitting some of their standard forms and documents to be reproduced as examples.

The photograph on the front cover shows a typical extension under construction. The photograph was supplied by Mr M. and Mrs A. Aldridge (of Cheshire) and Mr D. Glover (Building Contractor), and has been reproduced with their permission.

PART ONE: Introduction

1 A general guide to drawing the plan

GENERALLY

Figure 1.1 indicates the basic information that will be needed on most plans. The plan is not meant as a solution to all situations but to provide good grounding. If you produce a drawing something like this you are well on the way to getting approval to your scheme. For a typical house extension plan the following minimum details will probably be required by Planning and/or Building Control Departments (full details concerning Planning and Building Control are provided later):

(a) Plan of the existing (for a ground floor extension probably a plan of the ground floor will suffice) – minimum scale 1:100 (1:50 is preferred) with drainage details.
(b) Plan of proposed (ditto, showing existing house and extension) – minimum scale 1:100 (1:50 is preferred) with drainage details.
(c) Section through building – minimum scale 1:100 (1:50 is preferred).
(d) Existing rear, side and/or front elevation (as applicable) – minimum scale 1:100 (1:50 is preferred).
(e) Proposed rear side and/or front elevation (ditto) – minimum scale 1:100; (1:50 is preferred).
(f) Site location plan.
(g) Block plan (sometimes the site location plan will suffice).
(h) Full specification of materials to be used, cross-referenced to the drawing.

The plans should also show the following:

(a) The position of the ground levels.
(b) The position of the damp-proof courses and any other barriers to moisture.
(c) The position, form and dimensions of the foundations, walls, windows, floors, roofs and chimneys.
(d) The intended use of every room in the building(s).
(e) The provision made in the structure for protection against fire and for insulation against the transmission of heat and sound.

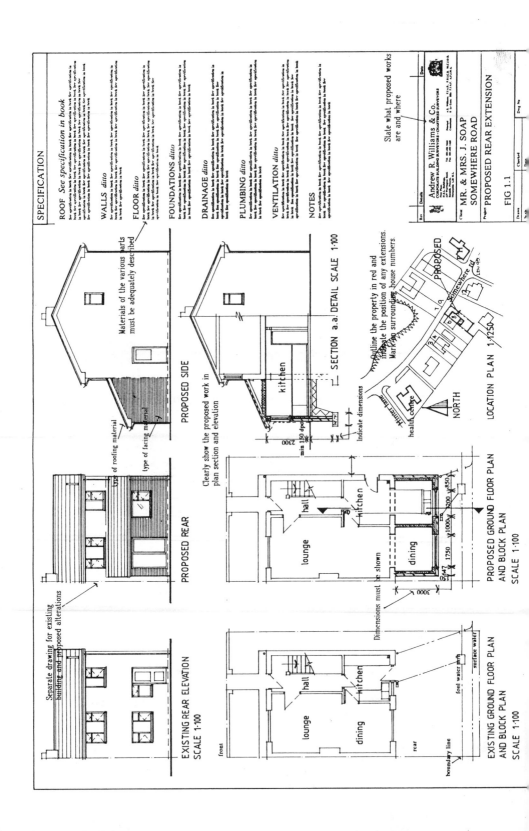

SPECIFICATION

ROOF *See specification in book*

WALLS *ditto*

FLOOR *ditto*

FOUNDATIONS *ditto*

DRAINAGE *ditto*

PLUMBING *ditto*

VENTILATION *ditto*

NOTES

Slate what proposed works are and where

Andrew R. Williams & Co.
CORPORATE BUILDING SURVEYORS · CHARTERED SURVEYORS

Client MR. & MRS. J. SOAP
SOMEWHERE ROAD

Project PROPOSED REAR EXTENSION

FIG 1.1

EXISTING REAR ELEVATION
SCALE 1:100

Separate drawing for existing building and proposed alterations

PROPOSED REAR

Type of roofing material

type of facing material

PROPOSED SIDE

Materials of the various parts must be adequately described

Clearly show the proposed work in plan section and elevation

kitchen

SECTION a.a. DETAIL SCALE 1:100

Indicate dimensions

min 150 dp

2300

Outline the property in red and indicate the position of any extensions.
Mark on surrounding house numbers.

PROPOSED

Somewhere rd

NORTH

health centre

LOCATION PLAN 1:1250

EXISTING GROUND FLOOR PLAN
AND BLOCK PLAN
SCALE 1:100

lounge

hall

kitchen

dining

front

rear

boundary line

foul water pipe

surface water

PROPOSED GROUND FLOOR PLAN
AND BLOCK PLAN
SCALE 1:100

lounge

hall

kitchen

dining

Dimensions must be shown

3000

1750

1000

1200

850

(f) The provision to be made for the drainage of the building or extension.
(g) Brief specification of materials and workmanship.
(h) The existing features of the site including any trees, outbuildings, those
     parts which will be demolished and be in sufficient detail to give a clear
     picture of any new building. Where existing and new works are shown on
     the same drawing, new work should be distinctively marked. In the past
     it was a common practice to colour wash drawings, but now this is not
     usually necessary as long as the intentions are clear and/or if hatching or
     some other device is used. The materials to be used in the external finish
     of walls and roofs and their colour should be indicated on the drawings
     (this is part of the specification on the drawing). On small works, the
     term 'to match' existing is normally sufficient when dealing with roofing
     and brickwork.

THE BLOCK PLAN/LOCATION PLAN

The location plan is normally drawn onto the main plan at a scale of 1:1250
and should include the north point. (Take a compass with you when you visit
the property to be extended and note north.) The location plan should also
show:

(a) The size and position of the building(s) and its (their) relationship to
     adjoining buildings.
(b) The width and position of every street adjoining the premises.
(c) The boundaries of the premises and the size and position of every other
     building and of every garden, yard and other open space within such
     boundaries.

At one time it was possible for anyone to trace this information from
Ordnance Survey sheets at the council offices. The officials were quite happy
to let you visit their offices armed with a small piece of tracing paper and a
pencil and extract the information needed. However, the Ordnance Survey
officials have now become far more strict regarding copyright. Except where
the council concerned relaxes the rules for owner applicants, nowadays, you
will either need to purchase a copy of the relevant sheet from an HMSO
agent, or make an accurate sketch of the surrounding area whilst measuring
the site and create your own location plan.
    In addition to a location plan, you should also provide a block plan.
(Sometimes the site location plan will suffice.)

SCALES GENERALLY

As indicated above, elevations and floor plans are normally to a scale of not
less than 1:100. It is recommended that scales of 1:50 be used wherever
possible as this makes the plans clearer, particularly when drawing sections.

Fig. 1.1 Plans of proposed rear extension.

LARGER PROJECTS

Obviously, the more ambitious the proposals, the larger number of plans, elevations and sections that will be required on each full drawing. If the extension or alterations affect two floors, then plans for both floors must be provided (proposed and existing). In theory, the plans should show every floor and roof of the building and a section of every storey of the building. In practice, for smaller projects, it is normally acceptable merely to show details of the floors being altered.

CHECKLIST OF ITEMS OFTEN MISSED OFF PLANS

In Appendix A I have included a short checklist of items that are often missed off plans or sometimes not addressed at all by the draughtsperson when preparing plans. The list is not intended to cover every eventuality and could doubtless be added to.

STANDARD SPECIFICATION

In Appendix B I have included a typical specification. It is very comprehensive and some might say it has an element of 'over-kill' about it. Because some items may not be applicable, it is usual to delete the inappropriate items when preparing a specific plan.

As it would be extremely tedious to have to hand-print this document onto every drawing, the simplest way of reproducing the standard specification on the drawings is by:

(a) Turning it into a booklet form and cross-referencing the main plan.
(b) Creating a master specification plan. Using this approach there are two alternatives:
   (i) You can make the master specification a full drawing in its own right, and copy it and send it in with the project drawing. This can be expensive as most systems of drawing reproduction are not cheap.
   (ii) You can affix the specification to the main drawing. The best system that I have found is by using Transtext. Transtext is a plastic film that will adhere to master drawings. Once attached, the Transtext patch does not normally show up on a well-produced print, and because it becomes a permanent part of the drawing, the information that it contains will not be mislaid. In my experience, it is this factor that makes most Building Control Departments prefer the Transtext method because incorporating the specification on the main drawing means that the details are unlikely to be overlooked once the work commences on site. By way of contrast, A4 enclosures in a small booklet have a nasty habit of disappearing once they reach site level –

out of sight means out of mind and so the builder 'does his own thing'.

(c) Storing the text on a computer or dedicated word processor for 'fine tuning' on each and every job. Using this approach there are two alternatives:

(i) modify and print out details as (a) above;

(ii) modify and print out the details and then photocopy onto Transtext.

THINKING IT THROUGH

Generally

Although it may seem fairly obvious, whether you are the builder/home owner or the surveyor who has been called in to prepare plans, it is essential from the very outset to decide exactly what is wanted. Before visiting a client, I often suggest that they make rough sketches of what they want to do to their property so that it is possible to swiftly discuss the practicality of a proposal once I arrive.

It has to be borne in mind that even though most homes lend themselves to extension, you need to weigh up many factors before finalizing the design. I have listed a few for consideration.

1. Does the improvement justify the expenditure?

It is possible to enlarge a property so much that the money invested is unlikely to ever be recovered if and when it is finally sold (e.g. someone enlarging a three-bedroom house into a six-bedroom house in an area that is principally composed of small properties is likely to be in this position.) I have actually been faced with situations where I have advised potential clients to move rather than extend, for exactly the reasons given above. Admittedly, I have had to forgo the odd commission, but I believe that honesty pays. I certainly would not wish to slip to the level of a high pressure salesperson and advise people to carry out a project that I believed to be foolhardy.

2. As well as the pluses are there minuses to the proposals?

Take as an example the family that want a dining room built onto their house. An extension of this nature that could only be reached via the existing kitchen is not as desirable as a dining room off a hallway. Kitchens can be untidy and smelly places whilst cooking is going on. Having to take guests through a working kitchen in order to reach the dining room might, in the long term, become an embarrassment. What if you wanted to impress your new boss or a highly critical relative? Having to drag them past pots and pans is far from ideal.

3. What will the effect be on the external appearance of the house?

If you drive around your home town for a short while, you will come across many examples of badly thought-out schemes. How about these for starters:

(a) The two-storey side extension to a normal two-storey house. The original house has a tiled roof but the extension has a flat roof. Admittedly, most Planning Departments nowadays reject applications like these, but some have been built. Very few people would describe this type of extension as beautiful, but it is amazing how many were built this way just to save money. (My office has recently designed a series of new tiled roofs to replace leaking flat roofs in our area when the owners suddenly realized that the original flat roof had been a false economy.) No doubt one or two planners let out a cheer as they were replaced.

(b) The loft conversion that is so dominant that it completely destroys the existing roof line.

(c) The extension built of badly matched bricks. (It saved money but the owner now lives to regret the penny pinching.)

4. What will be the effect on adjoining properties?

I have known several situations where good neighbours fell out merely because one ignored the wishes of another.

5. Is the proposed extension oversized and badly sited?

Badly designed extensions can result in loss of sunlight/outlook both for the instigator of the extension and the adjoining owners.

As people do still ignore obvious design considerations, some sort of control is needed, and that is the subject of the next chapter.

2   The need for control

WHO CARES WHAT THE NEIGHBOURS THINK?

Do you remember the newspaper articles and TV reports about the man who decided that a house was not really a home unless it had a very large model fish on the roof and a secondhand armoured fighting vehicle in the front garden? I gather from the media reports that his neighbours are not exactly pleased with his 'follies'.

COMMON SENSE IS NOT ALL THAT COMMON

There is an old saying that common sense is not all that common. Do you remember the newspaper story about the man who decided that it might be nice to have a basement under his house? You know the one, he laboured away all summer and as he was digging, he came across some obstructions in the ground (in the trade, we call them foundations), so he dug them up and the house, naturally enough, collapsed.

Knowing newspapers it might have been an apocryphal story, but I have seen members of the public doing some extremely stupid things over the years, either because they did not know any better, or because they had decided that the rules did not apply to them. Then there is the person who is doing something as a 'matter of principle'. Solicitors love them because as they say 'principles cost money'.

Unfortunately, there will always be some members of the 'it's-my-property-I'll-do-what-I-want-with-it' brigade in every area, and that is why the various rules and regulations become more strict as the years go by.

At this point, I am going to toss in a personal anecdote as a further illustration. Many years ago I was asked to prepare a set of plans for an extension on a property. As the enquiry was generated by a tradesman builder that I knew reasonably well, we set off together to visit the house in question. Once under way, Tony, the builder, explained that he had given the owner a budget quotation for the work based on a cost per square foot of floor area, and wanted to come with me to ensure that he had a good look around before firming up his price.

Once we arrived and had been invited in, Tony gave me a puzzled look and then whispered 'Funny house this!', and then went back to staring at the patch of new plaster on the ceiling. While Tony had been eyeing up the plaster, I had been weighing up a timber post in the middle of the living room. Then I realized that there was a sink on the far side of the living room. It was at that point that Tony drew my attention to the fresh plaster patch. This had, without any doubt, been the result of repair works following the removal of the chimney breast in the living room.

Being concerned about the structural integrity of the property, I said, 'Have you put in RSJs (rolled steel joists) to support the remaining chimney breast upstairs?' The owner looked at me and shook his head. It was at that point that I glanced around the rest of the house with more enlightenment. Only moments later, the truth dawned! This silly man had demolished all the load-bearing partition walls downstairs; the entire weight of the first floor was supported on the 4 in × 4 in (100 mm × 100 mm) wooden pole and the external walls. It was then that I warned the owner that I considered the whole property structurally unsound. I think that I upset him because the gentleman in question made it clear that if it had not been for his wife's insistence he would have built the new extension himself as well.

As far as he was concerned, he could see no reason for submitting plans to the council. It was all just 'red tape' designed to keep overpaid civil servants in the lap of luxury. He could not see anything wrong in his previous 'do-it-yourself' attempts. It was at that point that Tony and myself both came to the same conclusion: we did not want anything to do with this character.

However, what if you were the next door neighbour? Bearing in mind that this house was a mid-terraced property, would you like to live next door? When the house collapses, as it will in the fullness of time, and damages other houses, what will happen? Maybe someone might even be killed!

It is people like the man above, who do not know what they are doing, or who skimp to save money, that make controls so essential. Such examples also prove the need for competent designers and builders. People who have no knowledge of building construction and are not prepared to learn, should not dabble.

3 The Local Authority

THE POWERS THAT BE

As I have indicated in the previous chapter, some control is needed over 'development' and most sensible and informed people appreciate this, even though at times they become frustrated by the fact that the procedures seem to take so long.

It should also be obvious that, because of the actions of a few unthinking people in the past, there is now very little that can be done to a property without having to consult the Local Authority (the local council). Thinking people would probably also agree that if the regulations were relaxed that the 'fringe element' would return to their old ways. To some, having to conform to rules is seen as an infringement of their civil liberties, but as I have tried to indicate in the previous chapter, control over 'development' is no bad thing.

Although there are exceptions (which will be discussed later in more detail), no work should commence unless the necessary approvals have been obtained from the:

(a) Building Control Department (or an Approved Inspector)
(b) Planning Department/Planning Authority.

In some areas, such as conservation areas, additional approvals are also required. (Practically speaking, ignoring the time limits that both departments are supposed to work to, in my 'patch', it normally takes at least two months to go through all the procedures, but it can take longer.)

In the main, the Planning and Building Control Departments come under the control of the Local Authority. However, sometimes the planning function will be either partially or totally under the control of a non-elected body such as a New Town Development Agency. With privatization now being the new buzzword, councils have also lost some of their authority over building control now that Approved Inspectors have arrived on the scene. (I will describe the role of the Approved Inspector in a bit more detail later.)

Generally speaking, it would be unwise to commence work until plans have been drawn up, copies have been forwarded to the Building Control Department and Planning Department using the prescribed forms and both departments have given their written approval to the proposals.

WHAT IS THE DIFFERENCE BETWEEN PLANNING AND BUILDING CONTROL?

The two terms are of course a form of jargon. Like all professionals, Local Authority personnel have developed their own shorthand and they seem to surround themselves in their own form of 'newspeak'. This can cause confusion for lay people. Quite often, people telephone my office and say that someone from the Planning Department has told them something but once asked for more detail, it becomes obvious that it was probably a Building Control Officer who provided the information and *vice versa*. So, what is the difference? The easily understood answer is the Planning Department is interested in what the building looks like and Building Control wants to know if it is structurally sound.

To put it another way, the Planning Department will concern itself with such matters as the number of off-street car parking spaces that will be left once a house has been extended. It will be highly concerned if it becomes apparent that a proposed house or extension will occupy an excessively large amount of land available on the plot (the overdeveloped site), or if the proposals are likely to overcrowd a neighbouring property. Likewise, if a proposal is likely to 'overlook' a neighbour, the department will be concerned.

Beware! The criteria that Planning Departments apply to a problem might differ considerably from area to area. Most Local Authorities have their own guidelines and most will provide copies of their policies if asked to do so. As a glaring example of differing local policy decisions, in my 'patch' there is one Local Authority that only allows 3 m deep rear extensions to most semidetached and terraced properties. The adjacent authority allows 4 m deep rear extensions. As the boundary between the two authorities is often the centre line of a road, one should consider the difficulty that a consultant surveyor might have explaining to a prospective client who lives on the 'wrong side of the road' that he or she will not be allowed to do what a neighbour has done close by.

Building Control will concern itself with the structural stability of the proposal (i.e. are the walls strong enough to support the loads being placed on them), if the building is watertight and whether it will cause a danger to the present or future occupants or surrounding properties if there was ever a fire in the building.

PART TWO: More on Planning

4 Council planning policies

HOUSE EXTENSION POLICIES

Most councils issue guidelines for extensions to domestic properties in their area.

For simplicity, I have reproduced a typical householder's design policy (Figs. 4.1(a)–(d)). This has only been reproduced as a guide. When designing any building, it is essential to obtain the policy document issued by the Local Authority having control over the area in question. The policies differ from council to council. As the information is fairly self-explanatory, I would suggest that you study the document as a typical example of the sort of requirements that are normal.

Note: *In Chapter 5 I have provided information concerning Permitted Development. If the extension constitutes 'Permitted Development' (PD) then the council (from the Planning point of view) has virtually no control over the design. In other words, generally speaking, design policies can only be enforced if planning approval (permission) is needed.*

My advice is to check to see if you need planning permission first. It is often possible to get a situation arising where the local Planning Department would refuse a full planning application because their design criteria had not been implemented whilst simultaneously agreeing to the alteration under Permitted Development rules.

GENERAL NOTES ON HOUSEHOLDER EXTENSION POLICIES

The following comments are made based on what is normal in my area; in my experience, the policies of most Planning Departments cover the following:

(a) Harmonizing – most councils wish to ensure that all materials used in an extension match as closely as possible those of the existing structure. (They do not want it 'to stick out like a sore thumb'.)
(b) Overlooking neighbours – in order that the 'nosey neighbour' syndrome can be avoided, most Planning Departments have created guidelines and set down minimum distances between parallel windows at the front, side and back so that 'overlooking' is minimized.

Fig. 4.1(a)–(d) House Extensions – design guidance. (Reproduced with permission of Halton Borough Council.)

DESIGN GUIDANCE AND STANDARDS SUPPLEMENTARY TO POLICY H7

HOUSE EXTENSIONS

1. AIMS OF THE POLICY

1. To ensure that a domestic extension does not spoil the character of the original dwelling but relates closely to it and harmonises with the existing house in its scale, proportions, materials and appearance.

2. To preserve the essential character of the street and surrounding area.

3. To protect the amenity of occupiers of adjacent properties.

4. To avoid the creation of dangerous highway conditions.

5. To safeguard the provision of a reasonable private garden space.

2. APPLICATION OF THE POLICY

The definition of "house" in the policy includes bungalows but excludes flats or maisonettes.

"Extension" means all additions to the house whether attached or not, and includes garages.

The House Extension Policy also applies to:-

1. Houses which are listed buildings and buildings in Conservation areas.

2. Houses in the Green Belt.
However, due to the special characteristics of these areas, more stringent controls may need to be applied.

Exceptions may be considered for an extension to provide basic amenities or facilities at ground floor level for the disabled.

FOR ALL EXTENSIONS

3. DESIGN IN RELATION TO EXISTING DWELLING

An extension should relate closely to and harmonise with the existing building in its scale, proportions, materials and appearance. In particular:

i) it should normally form an unobtrusive and subordinate addition to the house.

ii) the external materials used should closely match those of the existing dwelling in colour, texture, profile and size or, according to the design, should provide a subordinate and sympathetic variation.

iii) on prominent elevations the problems of bonding of old with new brickwork on the same plane should be overcome by setting the extension back from the main wall of the dwelling.

iv) The roof of an extension should be pitched to match that of the existing dwelling. Flat roofs are not normally acceptable except where they are a feature of the original dwelling house.

v) the windows in an extension should line through with existing windows and should match their proportions, size and design.

4. EFFECT ON THE STREET SCENE AND THE CHARACTER OF THE AREA

Apart from its relationship to the existing house, an extension should not be visually detrimental to the existing character or appearance of the street scene or the surrounding area. In particular:

i) an extension should respect any regularity and width of spaces between existing houses and the visual effect of these spaces when significant in the street scene.

ii) an extension should respect any regularity in the distance between the road and the frontage walls of existing houses when this distance is a significant factor in the street scene.

iii) an extension at the rear of a dwelling should not be so extensive in relation to the size of the rear garden or yard that the enlarged house would constitute overdevelopment of the site which would be out of character with the general area.

iv) where a house is one of a group of similar appearance and significant in the street scene, the effect of an extension to that house on the appearance of the group, as well as the individual house, should be carefully considered.

5. AMENITY OF NEIGHBOURS

An extension should respect the existing standard of privacy and daylight experienced by neighbours, in particular:

i) where an extension will result in principal windows directly facing, or in principal windows directly facing a blank elevation, a minimum distance of 13m must be maintained.

ii) any new patio area or balcony at first floor level should not have the potential for an unacceptable degree of overlooking of any main window of a principal room in an adjacent house; nor for the direct sideways overlooking of a neighbouring private garden or yard.

Fig. 4.1(b)

6. HIGHWAY SAFETY

Extensions which prevent the parking of cars within the curtilage of the dwelling in accordance with the Council's car parking standard or which necessitate the use of the entire front garden for this purpose will not normally be acceptable.

An extension which incorporates a garage should have a minimum 5.5m distance between the front of the garage door and the public highway.

An extension should not be constructed in a position where it would interfere with an adequate standard of visibility for road users to the detriment of highway safety.

7. PORCHES/ FRONT EXTENSIONS

Where the front wall of a house conforms to a regular building line and this is a significant feature in the street scene, any single storey front extension or porch should not project by more than 1.5m from the main wall of the dwelling.

There should be a minimum distance of 4m between a front extension and the front boundary of the property.

8. SIDE EXTENSIONS
(a) Two Storey Side Extensions

In cases where a two storey side extension would produce a terracing affect the following shall apply:-

- a minimum gap of 1m shall be retained between the side wall of the extension and the plot boundary.

- and the extension shall be set back a minimum of 1.0m from the main front elevation of the existing dwelling.

(b) Two Storey Side Extensions on Semi-detached Properties where terracing is not an issue

In cases where a two storey side extension would unbalance the appearance of a pair of semi-detached properties the extension should normally be set back a minimum of 1 metre from the main front elevation of the existing dwelling and have a subordinate appearance.
Ideally it is preferable that **any two storey extension** is sited clear of the boundary and is designed with a pitched roof of the same type as the existing dwelling and with a lower ridge height than that of the main roof.

(c) Side Extensions on Corner Plots on Open Plan Estates

A side extension on a corner plot will not normally be permitted if the side wall of the extension is within 3m of any public footpath or highway verge.

9. REAR EXTENSIONS

a) Single Storey

i) An extension will not normally be allowed if it projects more than 4m from the main wall of the dwelling.

ii) A mono pitch roof will not normally be allowed where it would result in a sense of oppressiveness/ loss of light to a neighbouring dwelling (e.g. in extensions to terraced houses).

b) Two Storey

i) Two storey extensions along a shared boundary shall not project at first floor level by more than 2m.

Fig. 4.1(c)

ii) In any other case the following sizes shall be applied:-

Distance between extension & boundary	Maximum Projection at first floor level
1m	2.5m
2m	3m
3m or more	4m

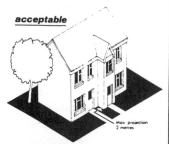

acceptable

Max projection
2 metres

10. ACCOMMODATION FOR THE EXTENDED FAMILY

The following guidance applies to proposals which enable elderly or dependant relatives with special needs to be accommodated within the existing family dwelling.

i) All extensions will be required to comply with preceding guidelines in terms of their siting, scale, design and external appearance.

ii) The additional accommodation should be attached to the existing property and be internally linked, rather than involving the construction of a separate building within the grounds. Where this is not practicable and it is proposed to convert and possibly extend an existing outbuilding, such as a garage, it will be expected that:

- the scale of accommodation to be provided is of a modest scale.

- sufficient space be available for parking in accordance with the Council's standards, clear of the highway without a significant loss of garden area.

iii) In all cases careful consideration will be given to the impact of proposals on neighbouring dwellings and any permission granted will be subject to a condition that the extension should be used for ancillary residential purposes in connection with the main dwelling and not as a separate unit of accommodation.

11. DORMER EXTENSIONS

Wherever possible dormer windows should be restricted to the rear of the dwelling in order to preserve the character of the street scene. This may not be so important where front dormers are already a common feature of other buildings in the street.

Side dormers on detached houses will be considered as front dormers, with additional regard paid to privacy and overlooking of adjoining properties.

Where dormers are on the front or side elevation of the dwelling or readily visible from a public place, their scale and design are particularly important and the following criteria will apply:

i) They should not normally exceed more than one third of the roof.

ii) They should not project above the ridge of the roof.

iii) Dormers which wrap around the side ridges of a hipped roof are not acceptable.

iv) The face of a dormer should be set back by a minimum of 1m behind the main front wall.

v) A dormer should not extend the full width of the roof, but should be set in from the side/party walls. Two smaller dormers may be better than one large one.

vi) Dormer windows should vertically line up with existing windows and match their style and proportions.

vii) Flat dormer roofs should be avoided where possible, unless considered more appropriate to the particular building or the street scene.

viii) Dormer cheeks should normally be clad in materials to match the existing roof.

unacceptable

12. PRIVATE GARDEN SPACE

Any extension to a dwelling will not normally be acceptable if it results in a rear usable garden space of less than 50 sq.m.

acceptable

Fig. 4.1(d)

(c) Minimum distance to rear boundary – as a further safeguard to privacy most councils require that when an extension is built, there should be a minimum distance between the rear of the extension and rear boundary of the property.

(d) Remaining garden space (overdevelopment) – some councils have a requirement that, once an extension is completed, a minimum area of free garden space is left at the rear. (On my 'patch', one council requires that at least 50 m^2 of free space is left.)

(e) Single-/two-storey extensions to the rear – most Planning Departments have requirements concerning how far an extension can project from the rear of a property when the proposal is fairly close to one or more side boundaries with a neighbour. Some have a set distance for both upper and lower storeys, others create formulas to be followed.

(f) Side extensions – most Planning Departments are concerned that side extensions should not create a 'terracing effect' (i.e. if two neighbours who own semidetached houses both extend them at ground and first floor level, then the block ends up looking like a terrace and has all the problems associated with terraced properties – lack of access to the rear, problems with sound transmission, people living 'cheek by jowl').

(g) Existing building lines – most Planning Departments try to restrict front extensions (other than very minor ones such as porches). Thus someone who seeks to obtain permission for a very dominant front extension is likely to receive a refusal.

(h) Flat roofs – flat roofs at second floor level are not normally looked upon with favour.

(i) Dormer extensions – where a loft is converted, and a dormer window is proposed, most Planning Departments will refuse an application if the new construction projects above the ridge line of the existing property.

(j) Garages and parking generally – to restrict on-street parking, most councils now require that adequate car parking is available once the proposed extension is completed. A proposal that completely uses up all available car parking space (or a substantial part of it) will normally be rejected. Most councils do not like garages to project in front to the foremost part of the house unless built with a porch. There is also usually a requirement that a driveway in front of the garage should be large enough to accommodate a car. (If you refer to Fig. 4.1(c) you will see that the council in question requires 5.5 m to be provided in front of the garage door.) This requirement is to ensure that the owner can fully pull off the road before opening the garage door. It makes good sense!

(k) Unsightly bonding on front elevations – in order to avoid cutting and bonding to brickwork being visible on front elevation most planners require a 'setback' of at least 112 mm where an extension is attached to the dwelling. Some councils require a far larger setback. If you refer to the Halton extension policy (Fig. 4.1(c)) you will note that a 1 m set back is required).

(l) Dominance – in my area, most councils require an extension to be 'an unobtrusive and subordinate' addition. If you refer to the Halton extension policy (Fig. 4.1(c) you will note that all the extensions deemed acceptable have main roof ridge lines lower than the original property and are subordinate in nature.

5 Is planning approval necessary?

PERMITTED DEVELOPMENT

As I have stressed the need for control in the first few chapters, the question 'is planning approval necessary?' may seem strange (as opposed to building control approval which is another matter). The simple fact is though, not all alterations require planning approval as long as certain criteria are met.

Although the Town and Country Planning Acts give Local Authorities the power to control 'Development', their powers are restricted in certain circumstances. If the proposals happen to fall within categories called **Permitted Development** (PD), then the Planning Department have to accept that the works can be carried out, whether they like it or not. (Remember, building control approval may still be required.) In England and Wales, what constitutes PD is defined in the Town and Country Planning General Development Order.

It may seem strange that there should be what is in effect a two-tier planning system, but it must be remembered that the restriction on the powers of the Planning Department is a common sense decision. If local council powers were total then the council offices would be flooded with all sorts of trivial applications which would quickly clog the system. No one would be able to do anything without needing to consult the local council. The PD system was designed to benefit both the Local Authorities and the general public.

For the designer and the home owner, the point at which a home extension ceases to be PD and requires a full planning application is very important.

NOTES ON PERMITTED DEVELOPMENT

Generally

As explained above, what constitutes PD is defined in a General Development Order (GDO) issued by the Secretary of State for the Environment and covers a wide range of 'Development'. The comments below only apply to domestic property and, like all things, GDOs are revised from time to time. (Beware – Article 4 of the GDO provides Local Authorities with the power to give a direction that Permitted Development rights can be suspended in certain circumstances. It is prudent to make enquiries with the Local Authority when proposing to deal with what is obviously a fine building to see if any restrictions are in force. This caution should also be exercised when dealing with works in National Parks, areas of outstanding beauty, conservation areas, buildings that are likely to be listed or where experience indicates that a previous planning consent has removed PD rights.)

Permitted Development is divided into many classes. (The 'rules' listed below are my rather simplistic interpretation of the GDO. For more information, you must obtain a copy of the current GDO and read it fully.) For convenience, I have divided PD as applicable to domestic alterations into the following parts.

(1) Porches.
(2) Dormer windows to lofts.
(3) Other extensions.

1. Porches

Porches can be built without planning approval (**note:** check with Local Authority if the house is in a National Park, an area of outstanding beauty, or a conservation area, or if it is listed, or if a previous planning approval may have removed PD rights) as long as:

(a) Ground floor area (measured externally) does not exceed 3 m².
(b) No part of the porch exceeds 3 m in height.
(c) No part of the house (or the porch when constructed) is closer than 2 m from a highway.

2. Dormer windows to lofts

Dormer windows to loft spaces can be constructed without planning approval (but check with Local Authority as for porches) as long as:
(a) The dormer would not project above the highest part of the roof (e.g. ridge line).
(b) The dormer does not face a road.
(c) Where a loft conversion would not add more than:
    (i) 40 m³ to a terraced house;
    (ii) 50 m³ to any other house.

3. Other extensions

Other extensions can be built without Planning Approval (but check with Local Authority as for porches) as long as:

A. *Terraced houses and houses on Article 1(5) Land*

Subject to the other rules regarding limitations on PD given below, terraced houses and houses on Article 1(5) Land, i.e. land within:

(a) A National Park
(b) Area of outstanding natural beauty
(c) Conservation area created under section 27 of the Town and Country Planning Act
(d) Area of natural beauty

can be extended by 50 m^3 or 10% of their floor area (whichever is greater) without planning permission subject to an upper limit of 115 m^3. (**Note:** where a proposed extension has a pitched (tiled or slated) roof, the cubic capacity of the roof has to be taken into account as well.)

B. *Other types of houses*

Subject to the rules regarding limitations on PD given below, detached and semidetached houses can be extended by 70 m^3 or 15% of their floor area (whichever is greater) without planning permission subject to an upper limit of 115 m^3. (**Note:** where a proposed extension has a pitched (tiled or slated) roof, the cubic capacity of the roof has to be taken into account as well.)

Limitations on PD for type 3 ('Other extensions')

The following limitations are applied to PD rights when dealing with 'type 3' extensions. (Speaking in very general terms, in practice, PD rights can only be effectively used for ground floor extensions which would not affect the existing main roof, but there are obviously always exceptions, as would be the case if you were dealing with a dwelling that was several storeys high.)

(a) No extension must be built higher than the existing house.
(b) The PD rules cannot be used in front garden situations (unless the front garden does not face a road or highway or the length of garden left after building the extension would exceed 20 m (approx. 66 ft.) (**Note:** the term 'Highway' does not just cover roads, it also covers public footpaths and bridleways. It presumably could also be interpreted as including an accessway sandwiched between rows of 'back to back' terraced houses.)
(c) The PD rules cannot be used where an extension faces a road or highway on any side of the house (Unless the length of garden left after building the extension would exceed 20 m (approx. 66 ft). **Note:** Take this rule into account when dealing with houses that have roads or highways front and back and/or side. As far as PD is concerned if a house has roads/highways on more than one side, then unless the length of the garden is big enough (20 m) planning permission must be obtained.)

(d) The PD rules cannot be used where an extension is within 2 m of a boundary **and** any part of the extension within that 2 m is over 4 m in height.

(e) The PD rules cannot be used where more than 50% of the existing garden area would be used up by the extension. (**Note:** when dealing with terraced houses with no front garden and only a small yard at the back this is a very real restriction.)

(f) The PD rules cannot be used for 'type 3' works where there would be an alteration to any part of the existing roof. (My interpretation of this rule would be that PD might not allowable if the proposal concerned the alteration of a bungalow and it is necessary to intersect with the existing pitched roof.)

Five metre rule

When calculating the cubic capacities for 'other types of houses' (3B above), the erection of a detached garage in the curtilage of the site is only treated as being an enlargement to the house if it is within 5 m of the house (or the proposed new extension to the house).

If it is possible to disregard the garage and its cubic capacity, then the owner of the property 'gets another bite of the cherry'.

FURTHER NOTES ON PD

Obviously some people might, if following the above rules without much thought, decide to extend their house by the Permitted Development limit one year and then try to do the same thing the year after and so on. **You cannot do this.** Once you have used up your permitted limit – that's it!

However, after saying that, there are always exceptions. The date 1st July 1948 is an important date to remember when dealing with PD. If your (or your client's) house has obviously been extended but it can be reasonably substantiated (say by examining the deeds of the property in question) that the extensions were carried out prior to 1st July 1948 (and no other extensions have been built after that date to use up the PD limit) then the PD allowances discussed above can be used to the full. In this case, the old extension is treated as being part of the original property.

One other important factor to take into account when using the benefits of PD, once you have a builder working on site, is that the builder is made aware of the rules that they must comply with. Some builders, if left to their own devices will build to suit themselves and if they exceed the stipulated limits and the Planning Department discover the discrepancies, they will be within their rights to demand a retrospective planning application. If for some reason they do not accept the alteration as being acceptable, the extension may need to be demolished or modified.

USING YOUR PD RIGHTS SENSIBLY

Generally

As indicated in Chapter 4, Planning Departments have design policies for their areas. In this chapter, I have indicated that these policies do not have to be applied when a proposal is deemed 'Permitted Development'.

However, I would not wish it to be thought that I am actively encouraging the populace as a whole to flout planning policies because in the main they have been created to protect the public (Chapter 2).

Purely as an example of how the system can be used sensibly, I intend to use another anecdote. A client that I once did work for wanted a 6 m long extension to his house. Under the policies applicable in the area in question he would only have been allowed to build a 4 m long extension. When I explained that a 6 m extension would contravene the council guidelines, this upset him because his neighbour had no objections to the extension and as proof he spoke to the man next door in my presence. He became even more upset when I explained that even though the neighbour might agree to the extra long extension, it was unlikely that the Planning Department would agree to it as it would create a precedent. When both parties told me they thought that that was a rather silly situation, I then explained that there might be a way around the impasse using their PD rights.

Playing the PD game

If you refer to Fig 5.1(a), you will see that I have made a sketch of my client's estate as originally constructed. All the houses are virtually identical. In Fig. 5.1(b), No. 3 Somewhere Road has built a 4 m long extension as per council policy, Nos 1 and 7 have done nothing whilst at no. 5, I proposed that part of the extension will be 6 m long and part 4 m long. Figure 5.1(c) explains how I achieved this. We constructed the extension in two phases. The 6 m extension was just under 70 m^3 and was built first. Once this had been built, we then applied for planning permission to build the 4 m section. It was quite legal!

So, as you can see, using your PD rights sensibly can help you in the following ways:

(a) As long as the property owner complies with the rules that govern PD, then he/she does not need to comply with local planning policy (e.g. if council policy only allows a 3 m rear extension but you want a larger extension, you can legally construct one as long as you obey say the 50 m^3/70 m^3 rules and all the riders).

Naturally, this is an anomaly that some Planning Officers dislike because once a proposal is permitted they are powerless to control the

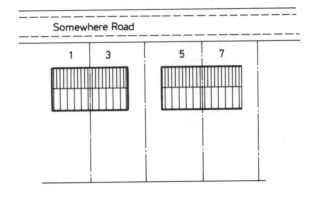

(a)

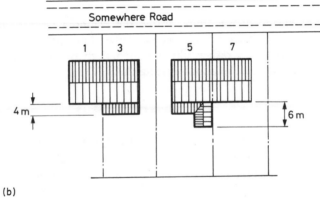

(b)

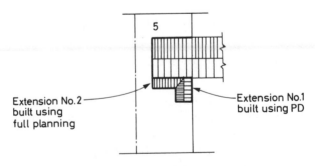

(c)

Fig. 5.1 Playing the PD game: (a) estate as originally built; (b) estate now; (c) how No. 5 used PD to his advantage.

works. (**Note**: you still have to comply with the requirements stipulated by Building Control.)

(b) If the client does not have to formally apply for planning permission then normally the extension can be built very quickly because the planning process normally takes at least two months to complete and a confirmation that PD is accepted only normally takes two to three weeks.

(c) Under recent legislation, Local Authorities charge for planning applications. If PD is applied, the fees (fee scale can be obtained from your local Planning Department) will not have to be paid.

However, **beware**. I never take the risk of informing a client that the extension is automatically exempt from the Planning Regulations. When I am dealing with an application which involves what I believe to be PD, I usually write to the planning department along these lines:

Bigtown Planning Department,
Municipal Offices,
Somewhere Road,
Hyville

Date

For the attention of Mr. Slowbend

Dear Sirs,

Proposed extension at 22 Bridge Street, Somewhere

Please find enclosed Plan No .... This indicates a tiled roofed extension, not exceeding 70 m^3 with a height not exceeding 4 m.

There have been no other extensions to the property (as far as we are aware).

We are of the opinion that this extension constitutes Permitted Development and would request your confirmation/comments in writing.

We would confirm that we have applied for building control approval under separate cover.

Yours faithfully,

A R Williams
FRICS FCIOB FBEng FIAS MIBC

By doing this you cover both yourself and your client.

Remember, if you or your client proceeds without obtaining written permission, there is always a chance, slim though it might be, that there is a reason why PD is not allowed. Obtaining a letter from the Local Authority confirming PD status proves that you did consult with the authorities before proceeding and ensures that there can be no 'comebacks' at a later date.

Let the Local Authority tell you that the scheme is acceptable under the PD rules.

6 Applying for planning permission

GENERALLY

You will realize from the last chapter that there are two alternative ways of dealing with the planning aspects of a proposed alteration/extension but I am going to go through the basics again.

Let us assume for the time being that you have prepared your plans and that you wish to submit them for approval. For a simple application there could be one of two alternatives, which are as follows:

(1) A letter to the Planning Department (enclosing a copy plan), requesting confirmation that the proposals are Permitted Development.
(2) A full planning application.

As I have covered the first alternative, let's assume that you need formal planning permission (approval).

THE FORMAL PLANNING APPLICATION

This chapter presumes that we are dealing with a fairly standard application and such things as tree preservation orders and the like do not apply. (This will be covered later.)

In order to obtain planning approval, your plans have to be submitted to the relevant Local Authority in your area together with a suitable application form. It is worth bearing in mind that once your application has been submitted to the Local Authority, anyone can ask to inspect the documents lodged. If you submit a scruffy set of documents, it will do you no credit.

Under normal circumstances, Local Authorities have their own specially printed planning application forms. Each Authority's forms vary slightly but the basic layout is the same. With the permission of Halton Borough Council I have reproduced a typical Householder planning application form (Fig. 6.1(a)–(b)) which is fairly straight forward to fill in. (**Note**: some authorities do not have Householder application forms. If this is the case with your authority, it will be necessary to fill in the full planning application form which is a little more complex.)

OFFICIAL USE ONLY - APP. NO. _____ DATE REC. _____

Town and Country Planning Act 1990

HOUSEHOLDER PLANNING
APPLICATION FORM

HALTON
BOROUGH COUNCIL
RUNCORN · WIDNES

USE THIS FORM FOR ANY PLANNING APPLICATION WHICH INVOLVES EXTENSIONS OR ALTERATIONS TO A DWELLING (INCLUDING THE ERECTION OF A GARAGE, CAR PORT, BUILDINGS, WALLS OR FENCES).

Send *THREE* copies of this form, together with *THREE* sets of plans, the appropriate certificate and fee to ;
Assistant Director (Planning) Municipal Building, Kingsway, Widnes. WA8 7QF

1 APPLICANT (BLOCK CAPITALS) PLEASE:

Name : *MR & MRS J. SOAP*
Address : *7, SOMEWHERE ROAD*
HALTON

Postcode : *H1 XY2*
Telephone : _____

2 AGENT (IF ANY) :

Name : *ANDREW R. WILLIAMS*
Address : *1437, WARRINGTON ROAD,*
RAINHILL

Postcode : *L35 4LL*
Telephone : *0151 426 9660*

3 FULL POSTAL ADDRESS OF THE APPLICATION SITE
(if different from the applicant's address)

7, SOMEWHERE ROAD, HALTON   H1 XY2

4 BRIEF DESCRIPTION OF PROPOSED WORKS
(stating number of storeys, position in relation to the existing building.)

SINGLE STOREY DINING AND KITCHEN EXTENSION

5 TYPE AND COLOUR OF MATERIALS TO BE USED
(Please show them on the detailed plans).

a   External wall facings   *TO MATCH EXISTING*
b   Roofing materials   *MARLEY MODERN TILES - COLOUR TO MATCH EXISTING*

6 EXISTING USE OF PROPERTY
Dwelling house (house/bungalow) [✓]   Flats [ ]   Other [ ]   (please specify)

7 ACCESS TO THE HIGHWAY
Do you intend to:
a   Construct a new access to a highway ?   *No*
b   Alter an existing access to a highway ?   *No*
   If you answer YES to a or b is it for vehicles and/or pedestrians ?
Please show details of any new or altered access on your detailed plan.

Fig. 6.1(a), (b) Householder planning application form. (Reproduced with permission of Halton Borough Council.)

8 **TREES**
Are any trees to be felled ?
If YES please show them on the detailed plans.

`NO`

9 **DEMOLITION OF WALLS AND OTHER BUILDINGS**
Do you intend to demolish any buildings ?
If YES please show them on the detailed plans.

`No`

10 **BUILDING REGULATION CONSENT** (most buildings need this sperate consent)
Have you submitted an application for Building Regulations Consent ?

`YES.`

11 **PLANS**
Please attach the following plans :
a Location plan, scale 1:1250 with site outlined in red.
b Detailed building plans, showing existing situation and the proposed works.
Please give plan numbers and list of information supplied ;

`FIG 1/1`

12 **FEES** (normally a cheque made out to Halton Borough Council in accordance with published scale fees.
Please attach the appropriate fee £ `80.00`

13 I/We apply for full planning permission for the proposals described in this application and the accompanying plans.

SIGNED `(signature)` DATE _____
On behalf of `MR & MRS J SOAP`

14 **CERTIFICATE OF OWNERSHIP**
Does all of the land on which you wish to build belong to you ? `YES`
If YES then :-
I certify that at the beginning of the period of 21 days ending with the date of the accompanying application, nobody except the applicant was an owner of any part of the land to which the application relates.None of the land to which the application relates constitutes or forms part of an agricultural holding. Section 65 Town and Country Planning Act 1990.

SIGNED `(signature)` DATE _____
On behalf of `MR & MRS J SOAP`

NOTE: *If the site or part of it is owned by someone else or is part of an agricultural holding you should ignore this certificate and instead complete an alternative certificate available from the planning office.*

15 **APPLICANTS CHECKLIST**
a 3 copies of the application form, signed and dated.
b At least 1 copy of the appropriate 'Certificate of Ownership' (Question 14), Signed and Dated
c 3 copies of location plan (edged red)
d 3 sets of detailed plans
e Appropriate fee

ENVIRONMENTAL SERVICES DEPARTMENT
Municipal Building, Kingsway, Widnes, Cheshire WA8 7QF. Telephone 051 424 2061 Fax: 051 495 1372

Fig. 6.1(b)

Parts 1 and **2** ask for the names and addresses of the applicant and agent. The applicant is the person who wishes to have the works carried out (e.g. the home owner). If you are working for someone (you have prepared the plans for someone other than yourself) and are looking after the submission, then you are the agent and you must insert your name and address in order that all correspondence will be directed to you. If you are not acting for someone (e.g. the application is for an extension on your own house), then there is no agent, so leave the box blank or cross it through. If you have a telephone number, provide it because if there are any queries, then someone can contact you easily and it helps to prevent delay.

Part 3 asks for the address applicable to the application. This address is not necessarily the same as the applicant's address. You or your client could be living at one address but own another property.

Part 4 then asks for brief particulars of the development. The word 'brief' should be noted. Do not write down a massive rambling description. If you are submitting plans for a 'single-storey kitchen extension to dwelling' then that is all that is required. If the Planning Officer needs amplification or the wording amending then he/she will contact you and say so. However, be careful. If your plans indicate a kitchen extension at the rear of the house and a porch at the front, make sure that you don't forget the porch. If you only ask for planning permission on the rear extension, then unless the case officer realizes that you have forgotten to mention the porch, you might end up only getting permission for one part of your scheme and have to re-apply for the part that you have forgotten to mention (and possibly pay an additional fee).

Part 5 on simple house plans the spaces can usually be filled in as follows:

(a) Facing brickwork to match existing.
(b) Tiles to match existing. In this particular case, as Marley Modern tiles are to be used, I have let the Planning Officer know the exact intentions.

Part 6 asks for current usage. As there is a tick box, it is easy to let the Planning Officer know that we are extending an existing dwelling house.

Part 7 asks about access to the highway. In the majority of cases both boxes will be filled in 'No'. But what if you need access to the highway? If your proposal is, say, for a new garage, and there is no pavement crossover in existence, it is essential that you ask for permission to put one in. If you need a pavement crossover but forgot to declare it on the form, you might end up having to re-apply for the pavement crossover separately and pay an additional fee.

Part 8 asks about trees. You must bear in mind that some trees have preservation orders on them and you cannot remove preserved trees without permission. Although it is unlikely that small trees are covered by an order, well established trees might well be. Show any tree affected by the proposals on your plan and answer the question truthfully.

Parts 9 and **10** are 'Yes/No' answers. With regards to Part 9, in most cases, the demolition of an old extension will not affect the application. The removal, for instance, of an old unsightly structure could well be a benefit not only to the applicant but also to the surrounding area. Refer also to Chapter 7 concerning conservation areas and the like. Part 10 is only really there to alert the applicant to the fact that building control approval might also be required.

Part 11 asks for a list of documents. In most cases, all that is required is to state the plan number on your drawing (this plan number could just be something simple such as Plan Number A1 or, if you have a numbering system, something more complex such as D/ARW/LF93/JN45B). However, you may also wish to enclose copies of Ordnance Survey sheets (if you have not drawn an extract of the OS sheet on your main plan), technical catalogues or letters from a neighbour or other relevant information. If you do, note them as well.

Part 12 asks for the Local Authority fee to be attached. At the time of writing, the planning fee for simple domestic extensions is £80.00. It is worth noting that if you or your client wanted to build two extensions and not just one, it is far more cost effective to submit both proposals together because the charge is per application, not per extension.

Note: a tip for agents/consultants/surveyors – I ensure that my clients pay the Local Authority fees themselves. I do not pay the Local Authority fees for them and then claim them back as a disbursement. Neither do I accept one cheque to cover my fees and the Local Authority fees. My reasons are twofold:

(a) If you pay the Local Authority fees, then you end up funding the client until planning approval has been obtained or until your client settles your account. In any business, cash flow is king and using your funds in this way does not make good business sense (and in the sad event of planning approval not being obtained, then some clients try to avoid paying their bills and you could end up 'holding the baby'.)

(b) If you accept Local Authority fee money and your own fees combined, part of that money is 'clients' money' (i.e. money held in trust, over which the surveyor has exclusive control and does not belong to him or her). The Royal Institution of Chartered Surveyors (RICS), for example, require a surveyor to keep a separate bank account for holding 'clients' money'. Not holding 'clients' money' saves a great deal of paperwork.

Part 13 merely asks you to sign the form. Do not forget to sign. The application cannot be accepted unless signed.

Part 14: the planning application is not complete without a certificate being included with it. Some Local Authorities include these on the planning forms (as with example), others use a separate sheet. Although the example form does not say so, it has a Certificate A printed on it.

The most usual certificates are Certificates A and B.

Certificate A (as on the example form) is used where no one other than the applicant has an interest in the land (e.g. the applicant is an owner occupier with a normal mortgage and his or her proposals do not interfere with the neighbours). When you sign this form you are certifying the following:

(a) That the applicant owns the land.
(b) That none of the land is part of an agricultural holding.

Certificate B is used when another party has an interest in the land (or part of the land). When you sign this form you are certifying the following:

(a) That you have notified all other owners of the land.
(b) That none of the land is part of an agricultural holding.

Note: Certificate B should be issued when a wall is built on a boundary. If ordinary footings are used the foundations will project into the neighbour's garden. This is legal if the neighbour agrees.

However – you cannot build on a neighbour's land without permission.

If you fill in Certificate B, a **Notice 1** (Fig. 6.2) has to be served on the person affected by the application. (This is compulsory if Certificate B is issued.)

A tip to save planning time: if you want to build on the boundary and you are using traditional foundations that project into the adjoining property, then get the neighbour to give you a letter saying that he/she has no objection to your proposals. You will still have to issue a Certificate B and Notice 1, but if you send a copy of the letter with your planning application, the council will know in advance that there will be no 'neighbour' problems. If the neighbour will not let you undermine his or her land, then you have to set your wall back minimum 150 mm from the boundary or use a tied footing. (Tied footings are described in Chapter 13).

Part 15 is a checklist of items to be submitted to the Local Authority. If you omit to provide the council with the relevant details there will be delays in processing the application.

Note also the requirement that the location plan on the drawings has to be edged in **red**. What this means is that you must colour the edge of the **plot of land** (not the extension) in question so that there can be no doubt concerning its location (Fig 1.1).

Town and Country Planning Act, 1990

Notice under Section 66 of application for planning permission

TO: *MR F BLOGGS* .... (owner of the undermentioned land)

5, SOMEWHERE ROAD, HALTON

TO: .... (agricultural tenant)

....

Insert address or location of proposed development.

Proposed development at ..... *7, SOMEWHERE ROAD, HALTON*

Insert name of applicant.

I GIVE NOTICE THAT ..... *MR J. SOAP*

Insert name of Council.

is applying to the ..... *HALTON BOROUGH* .... Council

Insert description of proposed development.

for planning permission to ..... *CONSTRUCT SINGLE STOREY DINING AND KITCHEN EXTENSION*

"Owner" means a person having a freehold interest or a leasehold interest the unexpired term of which was not less than 7 years.
If the application is for planning permission for the winning or working of minerals "owner" includes any person entitled to an interest in a mineral in the land (other than oil, gas, coal, gold or silver).
"Agricultural tenant" means a tenant of an agricultural holding.

Insert address of Council.

Any owner/agricultural tenant* of the land who wishes to make representations about this application should write to the Council at:-

PLANNING DEPT, MUNICIPAL BUILDINGS, KINGSWAY WIDNES, CHESHIRE WA8 7QF

within 21 days of the date of service/publication* of this notice.

Signed ....

*On behalf of ..... *MR J. SOAP*

Date ..... --/--/9--

* Delete where appropriate

Fig. 6.2 Notice 1.

7 Conservation areas/listed buildings/tree preservation orders

CONSERVATION AREAS

If you refer to Fig. 7.1(a)-(g) you will see a typical Conservation Area Policy. You will note that the Local Authority in question has made specific requirements concerning alteration works carried out in this conservation area. When dealing with work in a conservation area you must be prepared to make enquiries with the Local Authority regarding their policies.

In conservation areas, permitted development rules are sometimes suspended by an Article 4 Direction and planning permission is required for most alterations.

LISTED BUILDINGS

The effects of listing are similar to those created by work in conservation areas. When buildings are listed and graded, there is an obligation on the part of the Local Authority to notify the owner and occupier as soon as their house is placed on the list. It is unlikely therefore that an owner/client will be unaware that their house is listed. In general, if a property is listed it means that demolition, alteration and additions will be allowed only after proposals have been carefully examined and alterations or extensions must not deface the character of the original. The demolition or alteration of a listed building without prior permission is punishable by imprisonment and/or a heavy fine. Normally, when dealing with an alteration to a listed building a separate application in addition to planning application must be made to the Local Authority. I would suggest that you contact your Local Authority and ask for an application form for alteration to a listed building so that you can study the contents. In general, alterations to listed buildings have to be advertised in local newspapers giving details of the proposals and saying where the plans may be inspected. A notice giving similar details is usually put on the site.

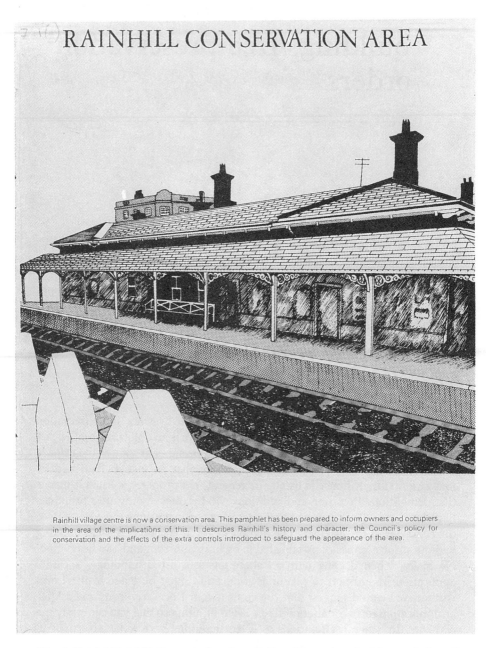

Rainhill village centre is now a conservation area. This pamphlet has been prepared to inform owners and occupiers in the area of the implications of this. It describes Rainhill's history and character, the Council's policy for conservation and the effects of the extra controls introduced to safeguard the appearance of the area.

Fig. 7.1(a)-(g) Rainhill Conservation Area Policy. (Reproduced with permission of St Helen's Borough Council.)

The History and Character of the Area

The village of Rainhill has developed over a long period of time. The name Rainhill is thought to derive from the Old English personal name of Regna or Regan. The earliest known reference is in 1190 when Richard de Eccleston granted to Alan the Clerk, his brother, the vill of Raynhull.

Until the late nineteenth century the village was an agricultural community. Some of the early recorded villagers reflect the nature of the community:

Edward Halsall, blacksmith, 1662;
Edward Whitlow, butcher, 1663;
Aeron Phythian, millwright, 1736;
Edward Parr, miller, 1713;
Thomas Hey, shoemaker, 1662;
George Smethers, stocking weaver, 1723;
Richard Glover, stonemason, 1620;
Thomas Glover, tanner, 1620;
Henry Garnett, weaver, 1635;

In addition there were the yeomen, husbandmen and labourers who worked on the farms.

The two most important events in the history of Rainhill were, first, the turnpiking of the highway from Liverpool to Warrington in the mid eighteenth century resulting in the improvement of the highway running through Rainhill from east to west and facilitating the development of the coaching service. Secondly, the construction of the Liverpool Manchester railway in the 1820's which helped its development later as a residential area. This railway is the first inter-city passenger railway not only in Britain but in the world.

In 1829 Rainhill was the venue for the locomotive trials to decide the type of engines which would operate on the railways. The engines had to traverse the level portion of the line from Rainhill Bridge towards Manchester and to complete a total run of 70 miles (the distance from Liverpool to Manchester and back) by 40 trips along the test length. Only the Rocket met the conditions and was declared the winner. These trials were of great importance in establishing the steam locomotive as the means of traction on the railways. The railway bridge which carries the Warrington Road over the railway is of particular importance. It was not possible to divert the road or railway to make a rectangular crossing and the bridge had to cross at an angle of 34°, forming a "skew" bridge. The bridge is important as an early example, if not the first, of its kind. The station too is of interest. The original station built in 1830 was called Kendricks Cross Station (a reference to Kendricks Cross which used to stand at the junction of Warrington Road and View Road) and was to the east of its present position. There was a level crossing joining what are now Victoria Road and Tasker Terrace thus maintaining the old highway from Cronton to Eccleston. The present station was built in the 1870's and retains its Victorian character.

The conservation area is concerned with the historic core of the village which developed along Warrington Road, either side of Kendricks Cross and the area between Warrington Road and the railway. This area is the present village centre which continues to fulfil its traditional function as a local centre containing the parish church, schools, shops, services, etc.

The character of the area is formed by the linear grouping of the buildings along Warrington Road. Although none of these buildings are outstanding, grouped together they form a pleasant environment. The area is visually enclosed by the rise of Warrington Road to the west to pass over the railway, by the curving tree-lined road to the east and by the buildings and large trees fronting onto Warrington Road. This enclosure together with the scale of the buildings gives the area a human scale which is worthy of conservation. The buildings also provide a link with the past, particularly the sandstone church and the old school buildings, giving a sense of continuity and stressing the development of the village from this small original core. The red sandstone seen in the old walls and buildings in fact bears witness to one of the oldest crafts in the township. A number of quarries were being worked in the area at one time but all have now ceased. Stone from View Road Quarry was used to build St. Ann's Church.

In order to protect the character of the area the Council intend to make Article 4 Directions. These have the effect of restricting specified classes of permitted development for which planning permission is not normally required, such as alterations to surrounding walls, building small extensions, garages, porches etc., changing windows and doors. A full list is given below. It should be stressed that it is not intended to prevent alterations and improvements from being carried out, but rather to ensure that they are carefully designed to be in character with the original. The Planning Section of the Council's Technical Services Department will give you advice prior to you making a formal application for planning permission should you be contemplating works on your property.

There is great scope for improvement within the conservation area. Most of these are of a minor, environmental nature, such as the landscaping and tidying up of vacant sites, landscaping of parking areas, painting and tidying up of some buildings, attention to walls and fences. Some of this work is the responsibility of the Local Authority, some is the responsibility of individual owners or occupiers. Proposals for the improvement of the area will be submitted to a public meeting to be held in Rainhill to enable all interested people and organisations to discuss the proposals.

Milestone on Bridge, Warrington Road (1829)

Fig. 7.1(b)

Warrington Road.

Article 4 Directions

The following classes of permitted development are to be restricted by the making of Article 4 Directions. Planning permission is therefore required for these works.

Class I – Development within the curtilage of a dwelling house.

(a) The enlargement, improvement or other alteration to a dwelling house.

(b) The erection or construction of a porch.

(c) The erection, construction or placing, and the maintenance, improvement or other alteration, within the curtilage of a dwelling house, of any building or enclosure required for a purpose incidental to the enjoyment of the dwelling house.

(d) The construction of a hardstanding for vehicles.

(e) The erection or placing of an oil storage tank for domestic heating.

Class II – Sundry minor operations

(a) The erection or construction of gates, fences, walls or other means of enclosure.

(b) The formation, laying out and construction of a means of access to a highway.

(c) The painting of the exterior of any building.

Class VIII – Development for industrial purposes

(a) Development of the following descriptions carried out by an industrial undertaker:

  (i) the provision, re-arrangement or replacement of private ways;

  (ii) the provision or re-arrangement of sewers, mains, pipes, cables or other apparatus;

  (iii) the installation or erection, by way of addition or replacement, of plant or machinery;

  (iv) the extension or alteration of buildings.

(b) The deposit of an industrial undertaker of waste material or refuse resulting from an industrial process.

Class IX – Repairs to unadopted streets and private ways.

The carrying out of works required for the maintenance or improvement of an unadopted street or private way.

Class XIII – Development by Local Authorities

(a) The erection or construction and the maintenance, improvement or other alteration by a local authority of:

  (i) such small ancillary buildings, works and equipment as are required on land belonging to or maintained by them;

  (ii) lamp standards, information kiosks, passenger shelters, public shelters and seats, telephone boxes, fire alarms, public drinking fountains, horse troughs, refuse bins or buckets, barriers for the control of persons waiting to enter public vehicles, and such similar structures or works as may be required in connection with the operation of any public service administered by them.

(b) The deposit by a local authority of work material or refuse on any land.

Class XIV – Development by local highway authorities

The carrying out by a local highway authority of any works required for or incidental to the maintenance or improvement of existing highways.

Class XVIII – Development by statutory undertakers

All development by statutory undertakers.

Fig. 7.1(c)

Warrington Road.

Policy

Conservation implies a long term view. It implies a responsibility to leave our historic environments at least as good as we found them. The need for Conservation can be argued in several ways: the links with our past provide us with a perspective against which we can evaluate our present actions; continuity in the environment enables us to have a sense of identity with familiar surroundings and this perhaps contributes to emotional and community stability; on a practical level conservation enables the careful management of limited resources.

Conservation cannot be considered in isolation. It should be seen in the context of other Borough Council functions. Designating an area as a conservation area does not mean that it will be preserved in its entirety as a museum piece. This would be unrealistic and would not take into account the other responsibilities of the Borough Council nor the social and economic forces over which the Borough Council has no control. An area must adapt to meet changing requirements and must be provided with those elements which allow it to continue to function. However, in designating a conservation area the Borough Council recognises the environmental quality of that area, and the importance of retaining this quality will be reflected in any future planning proposals and decisions. The Council will also take specific steps to protect, and where necessary improve, the environment within designated conservation areas. These steps are described below.

1 Development Control

The following general principles will be adopted for dealing with planning applications in conservation areas.

(a) Outline applications for planning permission will not normally be considered without sufficient details of the siting and design of a proposed building to show the proposal in the context of its surroundings.

(b) A high standard of design for new buildings and for the alteration or restoration of existing buildings will be expected. Any new building should be so designed as to harmonise in form, scale and materials with the area as a whole and with its immediate surroundings in particular. Care should be taken that the siting of proposed buildings is in sympathy with the pattern of existing frontages.

(c) Attention will be given to the proper planting and maintenance of trees. Developers will be encouraged to plant additional trees where appropriate as part of their development proposals as well as carrying out other landscaping work.

(d) It is important that materials used in a conservation area are sympathetic to that area and reflect its character. The use of cobbles, granite setts, brick pavers, stone flags etc. will be encouraged where appropriate and the Borough Council will keep a store of such materials. Careful attention will also be paid to the design and siting of street furniture and steps will be taken where possible to limit the intrusion of wires, posts, pipes etc.

2 Policies specific to individual Conservation Areas

Where necessary and when the opportunity arises detailed policies will be introduced for conservation and improvement in specific conservation areas. The introduction of these policies will depend on the availability of financial resources. Any proposals will be submitted for consideration to a public meeting in the areas to which they relate.

3 Residents in Conservation Areas

The Borough Council recognises that the success of any conservation policy depends upon the co-operation and the enthusiasm of residents, traders and property owners in the conservation areas. Even if the Borough Council was not strictly limited in manpower and in financial resources it would be vitally important that residents accept some of the responsibility for the appearance of their local environment. Much work has been done in the past by volunteer groups – clearing up of waste land, tree planting, landscaping, clearing of canals and ponds, restoration of buildings. The Borough Council will encourage any local initiative of this kind and give as much assistance as possible both in the organisation and implementation of such projects.

4 Conservation Advisory Group

The Council has established a Conservation Advisory Group consisting of representatives from Parish Councils, local and national amenity societies, local Chamber of Trade etc. The group is able to provide specialist and local knowledge and acts as a liaison body between the public and the Borough Council on Conservation matters.

The group may be contacted through its Chairman, c/o The Town Clerk, Town Hall, St. Helens.

Fig. 7.1(d)

Doors and Windows in the Area

Fig. 7.1(e)

Legislation

This Section summarises the relevant powers and duties of the Borough Council with regard to Conservation.

1 Designation of Conservation Areas

The designation of conservation areas is made under Section 277 of the 1971 Town and Country Planning Act as re-enacted by Section 1 of the Town and Country Amenities Act, 1974 which requires the Local Planning Authority to determine which parts of their area are of special architectural and historic interest, the character or appearance of which it is desirable to preserve or enhance and designate such areas as conservation areas.

In order to protect the character and appearance of conservation areas it is necessary to bring development under tighter control. The powers to do this, made available through various Acts of Parliament, are outlined below.

2 Demolition of Buildings

A building in a conservation area cannot be demolished without the consent of the Local Authority. An application to demolish a building may be made as a separate application or as part of an application for planning permission.
(Town and Country Planning Act, 1971, Section 55).

3 Urgent Repair of Unoccupied Buildings

If it appears to the Local Authority that any works are urgently required for the preservation of an unoccupied building in a conservation area and that it is important to preserve the building for the purpose of maintaining the character or appearance of the conservation area, then they may execute the works after giving the owner of the building not less than seven days' notice, in writing, of their intention to do so.

The Local Authority may give notice to the owner of the building requiring him to pay the expenses of any works executed.
(Town and Country Amenities Act, 1974, Section 5).

Late C19 House, Warrington Road.

Fig. 7.1(f)

4 Grants and Loans

(The term 'grant' in this section can be taken to include loans).

Grants are available in certain circumstances from both central government funds and from local authorities. They are always at the discretion of the body giving them and are naturally restricted by the amount of money available at a given time.

Exchequer Grants

The Secretary of State for the Environment has power to make grants for the repair and maintenance of buildings that are of outstanding architectural or historic interest. Comparatively few buildings qualify and so the scope for these grants is limited. (Historic Buildings and Ancient Monuments Act, 1935, Section 4). (Civic Amenities Act, 1967, Section 4).

Local Authority Grants

Local authorities have a wider scope. They can make grants for any building of architectural or historic interest and are not restricted to outstanding buildings. Again grants are payable for works of repair and maintenance. Grants are not payable for alterations or additions.
(Local Authorities (Historic Buildings) Act, 1962.
(Town and Country Planning Act, 1968, Section 58).

You may also be able to get a home improvement grant for improving or converting a building which is to be used as a dwelling.
(Housing Act, 1974).

5 Publicity for Planning Applications

Notice of any application for planning permission that would affect the appearance or the character of a conservation area is published in a local paper, stating where copies of the application and plans may be inspected by the public. The Local Authority will also, for not less than seven days, display on or near the land, a notice indicating the nature of the development in question. The application is not determined for 21 days from the date of publication of the notice. During this period the public may make representations to the Local Authority and any such representations are taken into account before the application is determined. This procedure may also apply to applications on land adjacent to a conservation area where the proposed development would affect the character of the conservation area. (Town and Country Planning Act, 1971, Section 28).

6 Protection of Trees

Anyone wishing to carry out work on a tree or to fell a tree in a conservation area must give the District Council six weeks' notice of his intention. During this time the Council may serve a tree preservation order on the tree. A public register of such notices is to be kept and the notices are valid for two years. Heavy fines are now imposed on any person who, in contravention of a tree preservation order, cuts down, uproots or wilfully damages a tree in such a manner as to be likely to destroy it.
(Town and Country Amenities Act, 1974, Section 8).

7 Control of Advertisements

The Local Authority can control the display of all advertisements in a conservation area.
(Town and Country Amenities Act, 1974, Section 3).

8 Town and Country Planning General Development Order, 1973

The Local Authority has the power to make the following directions subject to Ministerial confirmation for the protection of conservation areas:

(a) An Article 4 Direction. If it becomes necessary this direction allows the Local Authority to bring any class of "permitted" development (i.e. certain small scale alterations or additions) under specific development control in the whole or any part of a conservation area.

(b) An Article 5 Direction. Where outline planning permission ought not to be considered separately from questions of siting or design or external appearance of the building, access, or landscaping of the site, the Local Authority may within one month of the application notify the applicant that they are "unable to entertain it unless further details are submitted". The Local Authority must specify what further information they need for arriving at a decision. The applicant may either furnish the information or appeal to the Department of the Environment within six months of the notice.

Published by G. K. Perks, C.Eng., M.I.C.E., M.I.Mun.E.

Fig. 7.1(g)

TREES

Generally

The effects of tree roots and moisture extraction on foundations has been discussed elsewhere but you need to consider surrounding trees when submitting a planning application.

The law protects trees in several ways:

(a) In general the Local Authority have to be notified of any work to trees in a conservation area.
(b) Ditto if the tree is the subject of a tree preservation order.

Note: The Forestry Commission can become involved if a large number of trees are involved.

Trees in conservation areas

The general principle is that you need to write to the Local Authority if you intend to carry out work or remove a tree in a conservation area.

Tree preservation orders

A Local Authority may make a tree preservation order relating to any tree, group of trees or a belt of woodland, and these may be in fields, gardens or building sites. Hedges are not covered, but large trees growing in hedges could be. You cannot interfere with a preserved tree without express permission. This is one of the reasons why the planning application form asks if any trees are to be felled. Once the Planning Officer knows that a tree is going to be removed, he or she will check to ensure that the tree concerned has no order on it.

8 Other design aspects for consideration

GENERALLY

Despite what some of the planning guidelines say about extensions being made to look like subordinate structures, I believe that unless the requirements of the Planning Authority force you down the 'subordinate route' the objective when designing an extension, if at all possible, should be to make it look like part of the original structure and not an afterthought.

BRICKWORK DIMENSIONS

Figures 8.1 and 8.2 are extracts from Ibstock's Technical Notes and are based upon coordinating brick sizes of 225 × 112.5 × 75 m which includes 10 mm joints. When designing an extension to an existing property, the designer should wherever possible attempt to keep brick cutting to a minimum. Obviously in older properties the designer should do his or her best to provide a match as described below.

MATCHING MATERIALS

One of the problems that extensions create is that of matching new materials with the existing. For the most part, the matching of materials on very small domestic extensions does not come within the remit of the surveyor because he/she has carried out his/her function merely by stating on the plans that the 'facing bricks shall match the existing' or that 'the new roof tiles shall match the existing'. The client then passes this problem over to the builder.

But what if you are engaged to supervise the works, or the extension is for yourself? Then it is necessary to know how to find a good match.

There are several problems concerning matching brickwork and other materials with existing. I have nominated a few for consideration as follows:

(a) Unless the brick/roof tile is a local product or in fairly wide use in your area, it is sometimes difficult to identify the original brick/tile. The identification process can be made even more difficult because:

50 Other design aspects for consideration

IBSTOCK   TECHNICAL NOTES: CLAY BRICKS & BRICKWORK DIMENSIONS

Brickwork Dimensions

In the production of design and construction drawings for brickwork, the designer should think *Bricks* and minimise cutting by using brick dimensions wherever possible. An awareness that the architect understands the craftsman's problems, such as unnecessary, time consuming and costly cutting, encourages interest in the job and a better result. Brickwork dimensions tables are intended as an aid to architects and designers in the preparation of design and construction drawings.

Notes

(i) When the bricks to be used in piers or small width panels

of up to 4 bricks are all either above or below *work size*, adjustment of the pier or brick panel width should be considered to avoid unusually large or small width perpend joints.

(ii) It is helpful to the site to indicate on construction drawings the number of bricks which make up the dimension.

(iii) Further information on the placement of expansion joints will be found in Ibstock Technical Note – *Brickwork, Designing for Movement* and in BS 5628 Part III 1985 and *Code of Practice for Use of Masonry.*

(iv) Based upon bricks produced to BS3921 with *co-ordinating sizes* of 225 x 112.5 x 75 which includes 10mm joints.

Vertical brickwork courses dimensions tables
using 65mm bricks and 10mm joints

COURSES		COURSES		COURSES		COURSES	
1	75	31	2325	61	4575	91	6825
2	150	32	2400	62	4650	92	6900
3	225	33	2475	63	4725	93	6975
4	300	34	2550	64	4800	94	7050
5	375	35	2625	65	4875	95	7125
6	450	36	2700	66	4950	96	7200
7	525	37	2775	67	5025	97	7275
8	600	38	2850	68	5100	98	7350
9	675	39	2925	69	5175	99	7425
10	750	40	3000	70	5250	100	7500
11	825	41	3075	71	5325	101	7575
12	900	42	3150	72	5400	102	7650
13	975	43	3225	73	5475	103	7725
14	1050	44	3300	74	5550	104	7800
15	1125	45	3375	75	5625	105	7875
16	1200	46	3450	76	5700	106	7950
17	1275	47	3525	77	5775	107	8025
18	1350	48	3600	78	5850	108	8100
19	1425	49	3675	79	5925	109	8175
20	1500	50	3750	80	6000	110	8250
21	1575	51	3825	81	6075	111	8325
22	1650	52	3900	82	6150	112	8400
23	1725	53	3975	83	6225	113	8475
24	1800	54	4050	84	6300	114	8550
25	1875	55	4125	85	6375	115	8625
26	1950	56	4200	86	6450	116	8700
27	2025	57	4275	87	6525	117	8775
28	2100	58	4350	88	6600	118	8850
29	2175	59	4425	89	6675	119	8925
30	2250	60	4500	90	6750	120	9000

Fig. 8.1 Vertical brickwork dimensions. (Reproduced with permission of Ibstock Building Products Ltd.)

TECHNICAL NOTES: CLAY BRICKS & BRICKWORK DIMENSIONS　　　**IBSTOCK**

Horizontal brickwork dimensions tables
using 215 x 102.5 x 65mm bricks and 10mm joints

Number of bricks	OPENINGS	PANELS WITH OPPOSITE RETURN ENDS (co-ordinating size)	PIERS OR PANELS BETWEEN OPENINGS
½	122.5	112.5	102.5
1	235	225	215
1½	347.5	337.5	327.5
2	460	450	440
2½	572.5	562.5	552.5
3	685	675	665
3½	797.5	787.5	777.5
4	910	900	890
4½	1022.5	1012.5	1002.5
5	1135	1125	1115
5½	1247.5	1237.5	1227.5
6	1360	1350	1340
6½	1472.5	1462.5	1452.5
7	1585	1575	1565
7½	1697.5	1687.5	1677.5
8	1810	1800	1790
8½	1922.5	1912.5	1902.5
9	2035	2025	2015
9½	2147.5	2137.5	2127.5
10	2260	2250	2240
10½	2372.5	2362.5	2352.5
11	2485	2475	2465
11½	2597.5	2587.5	2577.5
12	2710	2700	2690
12½	2822.5	2812.5	2802.5
13	2935	2925	2915
13½	3047.5	3037.5	3027.5
14	3160	3150	3140
14½	3272.5	3262.5	3252.5
15	3385	3375	3365
15½	3497.5	3487.5	3477.5
16	3610	3600	3590
16½	3722.5	3712.5	3702.5
17	3835	3825	3815
17½	3947.5	3937.5	3927.5
18	4060	4050	4040
18½	4172.5	4162.5	4152.5
19	4285	4275	4265
19½	4397.5	4387.5	4377.5
20	4510	4500	4490
20½	4622.5	4612.5	4602.5
21	4735	4725	4715
21½	4847.5	4837.5	4827.5
22	4960	4950	4940
22½	5072.5	5062.5	5052.5
23	5185	5175	5165
23½	5297.5	5287.5	5277.5
24	5410	5400	5390
24½	5522.5	5512.5	5502.5
25	5635	5625	5615
25½	5747.5	5737.5	5727.5
26	5860	5850	5840
26½	5972.5	5962.5	5952.5

Consider expansion joint at 6mm centres in parapet and free standing walls

Number of bricks	OPENINGS	PANELS WITH OPPOSITE RETURN ENDS (co-ordinating size)	PIERS OR PANELS BETWEEN OPENINGS
27	6085	6075	6065
27½	6197.5	6187.5	6177.5
28	6310	6300	6290
28½	6422.5	6412.5	6402.5
29	6535	6525	6515
29½	6647.5	6637.5	6627.5
30	6760	6750	6740
30½	6872.5	6862.5	6852.5
31	6985	6975	6965
31½	7097.5	7087.5	7077.5
32	7210	7200	7190
32½	7322.5	7312.5	7302.5
33	7435	7425	7415
33½	7547.5	7537.5	7527.5
34	7660	7650	7640
34½	7772.5	7762.5	7752.5
35	7885	7875	7865
35½	7997.5	7987.5	7977.5
36	8110	8100	8090
36½	8222.5	8212.5	8202.5
37	8335	8325	8315
37½	8447.5	8437.5	8427.5
38	8560	8550	8540
38½	8672.5	8662.5	8652.5
39	8785	8775	8765
39½	8897.5	8887.5	8877.5
40	9010	9000	8990

Number of bricks	OPENINGS	PANELS WITH OPPOSITE RETURN ENDS (co-ordinating size)	PIERS OR PANELS BETWEEN OPENINGS
40½	9122.5	9112.5	9102.5
41	9235	9225	9215
41½	9347.5	9337.5	9327.5
42	9460	9450	9440
42½	9572.5	9562.5	9552.5
43	9685	9675	9665
43½	9797.5	9787.5	9777.5
44	9910	9900	9890
44½	10022.5	10012.5	10002.5
45	10135	10125	10115
45½	10247.5	10237.5	10227.5
46	10360	10350	10340
46½	10472.5	10462.5	10452.5
47	10585	10575	10565
47½	10697.5	10687.5	10677.5
48	10810	10800	10790
48½	10922.5	10912.5	10902.5
49	11035	11025	11015
49½	11147.5	11137.5	11127.5
50	11260	11250	11240
50½	11372.5	11362.5	11352.5
51	11485	11475	11465
51½	11597.5	11587.5	11577.5
52	11710	11700	11690
52½	11822.5	11812.5	11802.5
53	11935	11925	11915

Consider expansion joint at 12mm centres in all cases

Number of bricks	OPENINGS	PANELS WITH OPPOSITE RETURN ENDS (co-ordinating size)	PIERS OR PANELS BETWEEN OPENINGS
53½	12047.5	12037.5	12027.5
54	12160	12150	12140
54½	12272.5	12262.5	12252.5
55	12385	12375	12365
55½	12497.5	12487.5	12477.5
56	12610	12600	12590
56½	12722.5	12712.5	12702.5
57	12835	12825	12815
57½	12947.5	12937.5	12927.5
58	13060	13050	13040
58½	13172.5	13162.5	13152.5
59	13285	13275	13265
59½	13397.5	13387.5	13377.5
60	13510	13500	13490
60½	13622.5	13612.5	13602.5
61	13735	13725	13715
61½	13847.5	13837.5	13827.5
62	13960	13950	13940
62½	14072.5	14062.5	14052.5
63	14185	14175	14165
63½	14297.5	14287.5	14277.5
64	14410	14400	14390
64½	14522.5	14512.5	14502.5
65	14635	14625	14615
65½	14747.5	14737.5	14727.5
66	14860	14850	14840
66½	14972.5	14962.5	14952.5
67	15085	15075	15065
67½	15197.5	15187.5	15177.5
68	15310	15300	15290
68½	15422.5	15412.5	15402.5
69	15535	15525	15515
69½	15647.5	15637.5	15627.5
70	15760	15750	15740
70½	15872.5	15862.5	15852.5
71	15985	15975	15965
71½	16097.5	16087.5	16077.5
72	16210	16200	16190
72½	16322.5	16312.5	16302.5
73	16435	16425	16415
73½	16547.5	16537.5	16527.5
74	16660	16650	16640
74½	16772.5	16762.5	16752.5
75	16885	16875	16865
75½	16997.5	16987.5	16977.5
76	17110	17100	17090
76½	17222.5	17212.5	17202.5
77	17335	17325	17315
77½	17447.5	17437.5	17427.5
78	17560	17550	17540
78½	17672.5	17662.5	17652.5
79	17785	17775	17765
79½	17897.5	17887.5	17877.5
80	18010	18000	17990

Fig. 8.2 Horizontal brickwork dimensions. (Reproduced with permission of Ibstock Building Products Ltd.)

(i)   Colours fade and become stained over the years so even if it is possible to find the exact brick/tile match, the result may still be unsatisfactory.

(ii) Bricks/tiles made in different batches vary in colour.

(iii) Successive governments have used the building industry as an economic lever. Lack of demand during periods of 'belt tightening' has forced many manufacturers out of business. If a rival does not move in and buy up the rights to produce the range, then some products just simply become unavailable.

(iv) Standardization makes a product obsolete. This has happened with bricks. At one time all sorts of brick sizes were available (2 in, 3 in etc.).

(b) In order to make the brickwork match as close as possible, brick bond, toothing to existing and colour of mortars used also need consideration.

POSSIBLE SOLUTIONS

Matching tiles/slates

When dealing with a single-storey extension, the problem of matching is not quite so vital as when extending an existing roof. No doubt in your travels you will have seen examples of side extensions which have extended an existing roof and has left a visible junction between old and new. It is far better practice to salvage tiles/slates from one elevation and re-use them on the most prominent elevation to ensure that the roof covering is uniform. New tiles/slates should then be used on a less prominent elevation. Where dealing with, say, a single-storey extension where there are no abutting tiles and it is difficult to see both the older upper roof and the newer lower roof, unless you stand a long way back, it is usually possible to approximately match the tiles in colour only and the finished work not to look unsightly.

Matching bricks

If it is possible to get a good sample of an existing brick, then take it to a 'brick library'. You laugh . . . but there are such things. Whilst they are not common, large builder's merchants and specialist brick suppliers and some Local Authorities do have 'libraries'. Larger builder's merchants and suppliers are usually prepared to let you have sample bricks to take home/or to show to your client. It is also possible to inspect sample tiles in the same way.

Brick size

In days gone by, bricks came in a variety of sizes (2 in, 3 in etc). It is therefore essential to check the sizes of bricks whilst on site. (Bricks should be measured as size on face plus one mortar joint on the length and height.) What chance have you got of matching the new work to the old if the old walls are in obsolete three-inch bricks? The answer is none unless you can find a brick that gives a reasonable match in your brick library or can obtain 'secondhand' bricks.

Secondhand bricks and other salvaged materials

There are some suppliers that specialize in obtaining, cleaning and reselling secondhand bricks and other salvaged materials. If and when the need arises, appropriate enquiries should be made concerning the availability of recycled products. Correct use of secondhand materials can overcome the problem of matching old to new. However, when using secondhand bricks in particular, it would be worth checking with the proposed supplier to see if they have been dipped or treated in any way. Untreated secondhand bricks can introduce problems, such as dry rot, into the property to be extended.

Matching mortar

It is possible to obtain premixed coloured mortar; several firms manufacture it (e.g. Tilcon, RMC). The mortar comes in a wide variety of shades and can usually be made to imitate the mortar colouring of the existing house. Most mortar suppliers are prepared to supply a set of small samples which give an indication of every colour that they make. Once a near match has been established, these companies will then usually supply a small pack of mortar so that a 'sample panel' can be constructed for client's approval. The 'rough stuff' that is supplied, however, has to be mixed with cement before use. Ask the supplier to advise on the amount of cement required.

Matching bond

There are several types of bond for brickwork (Figs 8.3 to 8.6). As cavity walls are now the most common form of construction, stretcher bond is normally used in cavity walls. But what if the existing house wall is solid construction and built in English bond or Flemish bond? The way around this problem is to use 'snap headers' on the outer skin. In other words your external cavity wall will still only be half a brick thick (112 mm), but the bricklayer will chop the headers in half to imitate the old brickwork bonding that you need to match.

Vertical abutments

Where new walls abut old walls the brickwork should be cut, toothed and bonded to the existing, in other words, pockets cut in the old brickwork and

**The external appearance
shown as all stretchers**

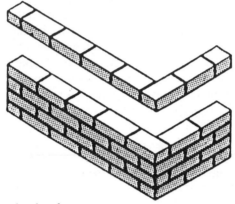

Fig. 8.3 Stretcher bond.

**The external appearance
shows alternate layers of
headers and stretchers**

Fig. 8.4 English bond.

The external appearance shows mainly stretchers with a row of headers every three, five or seven courses

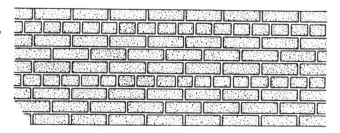

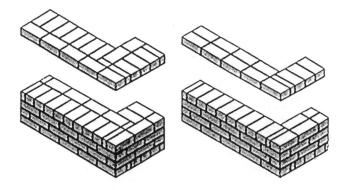

Fig. 8.5 English garden wall bond.

The external appearance shows alternate headers and stretchers in each course

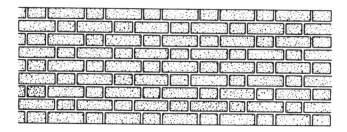

Fig. 8.6 Flemish bond.

the new bricks let in. If you refer to Fig. 8.7, you will note that the cutting and toothing can remain permanently on display unless carried out with care. Figure 8.8 shows a way to avoid the cutting, toothing and bonding showing at corners by setting the brickwork back by half a course. If you do not wish to use the solution shown in Figure 8.8, the only way of totally avoiding cutting, toothing and bonding is by using patent jointing strips such as 'Furfix' (or

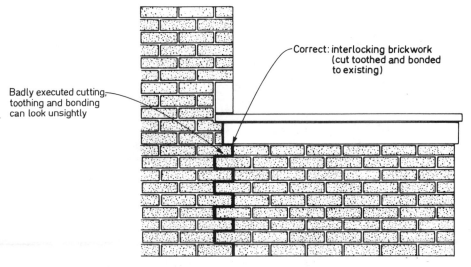

Correct: interlocking brickwork (cut toothed and bonded to existing)

Badly executed cutting, toothing and bonding can look unsightly

Fig. 8.7 Vertical abutments: cut, tooth and bond.

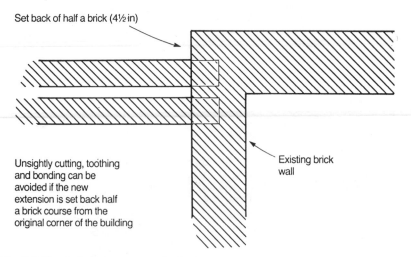

Set back of half a brick (4½ in)

Unsightly cutting, toothing and bonding can be avoided if the new extension is set back half a brick course from the original corner of the building

Existing brick wall

Fig. 8.8 Vertical abutments: set back at corner.

Vertical abutments.   Where new walls abut old walls the brickwork should be cut, toothed and bonded to the existing. In other words, pockets cut in the old brickwork and the new bricks let in

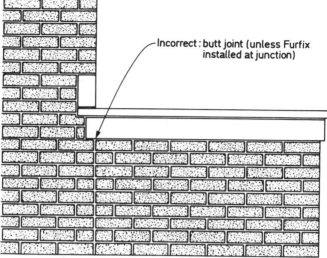

Incorrect : butt joint (unless Furfix installed at junction)

Fig. 8.9 Vertical abutments: butt joint.

other similar system) which is bolted to the existing wall and has steel 'teeth' that bed into the mortar joints of the new brickwork. This modern approach to jointing new to old has its advantages, but the manufacturer's instructions should be followed when using any product such as this, in order to provide a watertight joint. On the down side, the problem with using 'Furfix' or similar is that a vertical joint can be seen running down the wall face (Fig. 8.9). As this vertical jointing could look identical to 'butt jointing' (which is bad building practice), the form of construction could mislead another surveyor when visiting the house to carry out a building survey for a future buyer.

Render finishes

Where a house is totally rendered or a combination of brick and render, try to line up with the original coursing. If the old render terminates in a bellmouth (Figure 12.1), then put one on the new render to match in.

Lining up fascia boards and windows

Where possible, line up all horizontals. There is nothing worse than a house that has window heads at irregular heights in relation to one another. Fascia boards should also be lined around if possible because it will help to create the illusion that the extension has always been there. The continuous lines help to 'weld' the extended property together.

SITE OBSTRUCTIONS

Although I will be discussing site visits in more detail later in the book, whether you are the owner or an outside consultant, you should be on the lookout for site obstructions. Here are a few examples of typical obstructions to the proposed development.

Trees

Forgetting for the moment possible tree preservation orders, any tree growing near to a house or extension can endanger the stability of the property (Fig. 8.10).

The *NHBC Standards* (Volume 1) chapter 4.2 now covers the subject of trees in a scientific way and considers various types of trees and relates their effects to various types of soil conditions. If you refer to Figs 8.11 and 8.12, which are extracts from *NHBC Standards* 'Building Near Trees' you will note some of their requirements when building on high shrinkability soils.

As an example, using these rules, if we were dealing with a poplar tree (anticipated mature height is stated as being likely to be 28 m) growing on a high shrinkable soil, then the distance from the centre of the tree to the house

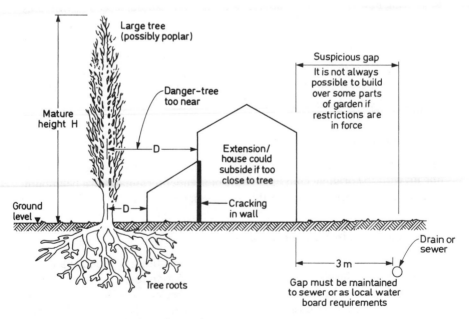

Fig. 8.10 Trees and other obstructions.

APPENDIX 4.2-B

4.2 Building near trees

Guidance

Water demand and mature height of trees

Broad leaved trees			Conifers			Orchard trees (take as broad leaved)		
Water demand	Species	Mature height [m]	Water demand	Species	Mature height [m]	Water demand	Species	Mature height [m]
High	Elm		High	Cypress				
	English	24		Lawson's	18			
	Wheatley	22		Leyland	20			
	Wych	18		Monterey	20			
	Eucalyptus	18						
	Oak							
	English	20						
	Holm	16						
	Red	24						
	Turkey	24						
	Poplar							
	Hybrid black	28						
	Lombardy	25						
	Willow							
	Crack	24						
	Weeping	16						
	White	24						
Moderate	Acacia False	18	Moderate	Cedar	20	Moderate	Apple	9
	Alder	18		Douglas fir	20		Cherry	15
	Ash	23		Pine	20		Pear	12
	Bay Laurel	10		Spruce	18		Plum	10
	Blackthorn	8		Wellingtonia	30			
	Cherry			Yew	12			
	Japanese	9						
	Laurel	8						
	Wild	17						
	Hawthorn	10						
	Honey locust	14	**Note**					
	Hornbeam	17	1 Where hedgerows contain trees, their effects should be					
	Horse chestnut	20	assessed separately. In hedgerows, the height of species					
	Laburnum	12	likely to have the greatest effect should be used.					
	Lime	22	2 Within the classes of water demand, species are listed					
	Maple		alphabetically; the order does not signify any gradation in					
	Japanese	8	water demand.					
	Norway	18	3 When the precise species is unknown the greatest height					
	Mountain ash	11	and highest water demand should be assumed.					
	Plane	26	4 Further information regarding trees may be obtained from					
	Sycamore	22	the Arboricultural Association or the Arboricultural Advisory					
	Tree of heaven	20	and Information Service (see Appendix 4.2-G).					
	Walnut	18						
	Whitebeam	12						
Low	Beech	20						
	Birch	14						
	Holly	12						
	Magnolia	9						
	Mulberry	9						

Fig. 8.11 'Building near trees', *NHBC Standards*, Appendix 4.2-B. (Reproduced with permission of NHBC.)

APPENDIX 4.2-H

| 4.2 | Building near trees |

Guidance

Foundation depth tables (continued)

TABLE 12

HIGH SHRINKAGE SOILS Plasticity Index greater than 40%

HIGH water demand BROAD LEAF trees including:-   Figures are heights in metres

Elm		Oak		Poplar		Willow	
English	24	English	20	Hybrid black	28	Crack	24
Wheatley	22	Holm	16	Lombardy	25	Weeping	16
Wych	18	Red	24			White	24
Eucalyptus	18	Turkey	24				

Foundation depths (m)

Distance from centre of tree or hedgerow to face of foundation (m)

Mature tree height (m)	1.00	2.00	3.00	4.00	5.00	6.00	7.00	8.00	9.00	10.00	11.00	12.00	13.00	14.00	15.00	16.00	17.00	18.00	19.00
8	3.25	3.00	2.75	2.50	2.25	2.00	1.75	1.50	1.25	1.00	1.00	1.00							
10	3.30	3.10	2.90	2.70	2.50	2.30	2.10	1.90	1.70	1.50	1.30	1.10	1.00	1.00					
12	3.33	3.17	3.00	2.83	2.67	2.50	2.33	2.17	2.00	1.83	1.67	1.50	1.33	1.17	1.00	1.00	1.00		
14	3.36	3.21	3.07	2.93	2.79	2.64	2.50	2.36	2.21	2.07	1.93	1.79	1.64	1.50	1.36	1.21	1.07	1.00	1.00
16	3.38	3.25	3.13	3.00	2.88	2.75	2.63	2.50	2.38	2.25	2.13	2.00	1.88	1.75	1.63	1.50	1.38	1.25	1.13
18	3.39	3.28	3.17	3.06	2.94	2.83	2.72	2.61	2.50	2.39	2.28	2.17	2.06	1.94	1.83	1.72	1.61	1.50	1.39
20	3.40	3.30	3.20	3.10	3.00	2.90	2.80	2.70	2.60	2.50	2.40	2.30	2.20	2.10	2.00	1.90	1.80	1.70	1.60
22	3.41	3.32	3.23	3.14	3.05	2.95	2.86	2.77	2.68	2.59	2.50	2.41	2.32	2.23	2.14	2.05	1.95	1.86	1.77
24	3.42	3.33	3.25	3.17	3.08	3.00	2.92	2.83	2.75	2.67	2.58	2.50	2.42	2.33	2.25	2.17	2.08	2.00	1.92
26	3.42	3.35	3.27	3.19	3.12	3.04	2.96	2.88	2.81	2.73	2.65	2.58	2.50	2.42	2.35	2.27	2.19	2.12	2.04
28	3.43	3.36	3.29	3.21	3.14	3.07	3.00	2.93	2.86	2.79	2.71	2.64	2.57	2.50	2.43	2.36	2.29	2.21	2.14
30	3.43	3.37	3.30	3.23	3.17	3.10	3.03	2.97	2.90	2.83	2.77	2.70	2.63	2.57	2.50	2.43	2.37	2.30	2.23

Foundation depths in excess of 2.5m should be Engineer-designed taking account of the likely movement of the soil on the foundations and sub structure.

TABLE 13

HIGH SHRINKAGE SOILS Plasticity Index greater than 40%

HIGH water demand CONIFEROUS trees including:-   Figures are heights in metres

Cypress	
Lawson's	18
Leyland	20
Monterey	20

Foundation depths (m)

Distance from centre of tree or hedgerow to face of foundation (m)

Mature tree height (m)	1.00	2.00	3.00	4.00	5.00	6.00	7.00	8.00	9.00	10.00	11.00	12.00	13.00	14.00	15.00	16.00	17.00	18.00	19.00
8	2.98	2.46	1.94	1.42	1.00														
10	3.08	2.67	2.25	1.83	1.42	1.00	1.00												
12	3.15	2.81	2.46	2.11	1.76	1.42	1.07	1.00											
14	3.20	2.90	2.61	2.31	2.01	1.71	1.42	1.12	1.00										
16	3.24	2.98	2.72	2.46	2.20	1.94	1.68	1.42	1.16	1.00									
18	3.27	3.04	2.81	2.57	2.34	2.11	1.88	1.65	1.42	1.18	1.00								
20	3.29	3.08	2.87	2.67	2.46	2.25	2.04	1.83	1.62	1.42	1.21	1.00	1.00						
22	3.31	3.12	2.93	2.74	2.55	2.36	2.17	1.98	1.80	1.61	1.42	1.23	1.04	1.00					
24	3.33	3.15	2.98	2.81	2.63	2.46	2.28	2.11	1.94	1.76	1.59	1.42	1.24	1.07	1.00				
26	3.35	3.18	3.02	2.86	2.70	2.54	2.38	2.22	2.06	1.90	1.74	1.58	1.42	1.26	1.10	1.00			
28	3.35	3.20	3.05	2.90	2.76	2.61	2.46	2.31	2.16	2.01	1.86	1.71	1.57	1.42	1.27	1.12	1.00		
30	3.36	3.22	3.08	2.94	2.81	2.67	2.53	2.39	2.25	2.11	1.97	1.83	1.69	1.56	1.42	1.28	1.14	1.00	

Foundation depths in excess of 2.5m should be Engineer-designed taking account of the likely movement of the soil on the foundations and sub structure.

Fig. 8.12(a), (b) 'Building near trees', *NHBC Standards*, Appendix 4.2-H. (Reproduced with permission of NHBC.)

APPENDIX 4.2-H

Building near trees | 4.2

Guidance

NOTES

1   Determine foundation depth using mature height of tree and its distance to face of foundation.

2   Interpolation between figures is permitted. Depths can be reduced by 0.05m (50mm) for every 50 miles north and west of Central London (see Appendix 4.2-D). It is not permissible to reduce the depth to less than the minimum given in the tables.

3   If the foundation depth occurs within the shading shown thus [＿＿＿＿] special precautions may be required to prevent excessive damage to tree roots (see Appendix 4.2-F).

4   If the combination of mature tree height and distance to face of foundation gives a result in the shaded area shown thus [░░░░░] then the foundation is outside the zone of influence for that tree and the minimum foundation depth may be used as indicated on each table.

5   Check that all the requirements of Chapter 4.2 are met - see Step 7 at the start of this Appendix.

Foundation depths (m)

Distance from centre of tree or hedgerow to face of foundation (m)

20.00	21.00	22.00	23.00	24.00	25.00	26.00	27.00	28.00	29.00	30.00	31.00	32.00	33.00	34.00	35.00	36.00	37.00	38.00

Use minimum foundation depth of 1m

1.00	1.00	1.00																
1.28	1.17	1.06	1.00	1.00														
1.50	1.40	1.30	1.20	1.10	1.00	1.00	1.00											
1.68	1.59	1.50	1.41	1.32	1.23	1.14	1.05	1.00	1.00									
1.83	1.75	1.67	1.58	1.50	1.42	1.33	1.25	1.17	1.08	1.00	1.00	1.00						
1.96	1.88	1.81	1.73	1.65	1.58	1.50	1.42	1.35	1.27	1.19	1.12	1.04	1.00	1.00				
2.07	2.00	1.93	1.86	1.79	1.71	1.64	1.57	1.50	1.43	1.36	1.29	1.21	1.14	1.07	1.00	1.00	1.00	
2.17	2.10	2.03	1.97	1.90	1.83	1.77	1.70	1.63	1.57	1.50	1.43	1.37	1.30	1.23	1.17	1.10	1.03	1.00

Foundation depths (m)

Distance from centre of tree or hedgerow to face of foundation (m)

20.00	21.00	22.00	23.00	24.00	25.00	26.00	27.00	28.00	29.00	30.00	31.00	32.00	33.00	34.00	35.00	36.00	37.00	38.00

Use minimum foundation depth of 1m

Fig. 8.12(b)

has to be at least 35 m before normal footings 1 m deep can be used. At 14 m away, then the depth has to be 2.50 m. Closer than that and the foundations need to be engineer designed and special precautions must also be made to protect the foundations against excessive root damage.

Sewers and buried mains cables

On some housing estates you may note apparently unaccountably large gaps between your house/prospective client's property and the next house, or that you or your client has an unusually large garden. Beware! Ask questions! Does anyone know of any sewer or buried cables in the vicinity? Builders never leave large strips of land undeveloped out of the goodness of their hearts. Prime building land costs a great deal of money and they will naturally try to build as many houses on a plot as they can. If you/your prospective client thinks that there might be a main sewer or electric cable in the vicinity . . . check! (Fig. 8.10). In the first instance it would be advisable to contact the Main Drainage Section at your local council. (They may have a different title at your council but do not be put off – you want to speak to the person/section that deals with sewers.)

Note: I am using the term 'sewer' as opposed to 'drain' quite deliberately. Sewers are larger than the normal 100/150 mm drainage that serves the house itself and are normally owned and maintained by the council or the water company in your area. In my experience, most sewers owned by the water company have restrictions placed upon them, and property owners are not normally allowed to build near to a sewer. In my area, the general rule is that no permanent structure shall be built closer than 3 m from the edge of a sewer, but in one recent plan that I prepared a 4 m gap was required. If you don't make some enquiries before you start preparing plans then both you and your client could be in for a disappointment. In the worst case, you could find that your plans will be rejected totally and that no amount of amendment will make your proposals acceptable. If the obstruction just turns out to be a deep drain, then the only problem is ensuring that the foundations of the house will not subside into it in the event of a failure. This can be resolved by concrete encasing of the drain or by putting in deeper foundations to the extension. This aspect of design is covered later chapters. Large electric cables installed by the local electricity company can also cause similar problems to sewers. If you believe that there might be a high voltage cable, water main, gas main or any other obstruction buried on the site, you can find out the truth by contacting your local electricity, gas or water companies. Normally, they will mark up a plan of their services and send it to you, without charge. The only snag is that large red disclaimer that they have stamped on the plan saying that the service position is only approximate to plus or minus six feet. If the service is very close to the proposal, you could still have a problem.

The unco-operative next-door neighbour

With most small extensions, the owner of the property to be extended wishes to maximize his or her benefits by building up to the boundary of his/her property. Normally neighbours will not object to this because they will be on friendly terms with either you or your client and will accept that the neighbours desire to better themselves as a reasonable aspiration. Unfortunately this is not always the case. There are two simple solutions to the problem of the awkward neighbour. The first is shown in Fig. 8.13(a). Here the new extension is set back 150 mm which means that the footing underground will not go under the neighbour's garden. (Foundations normally have a 150 mm spread.) The second solution is to use a tied footing

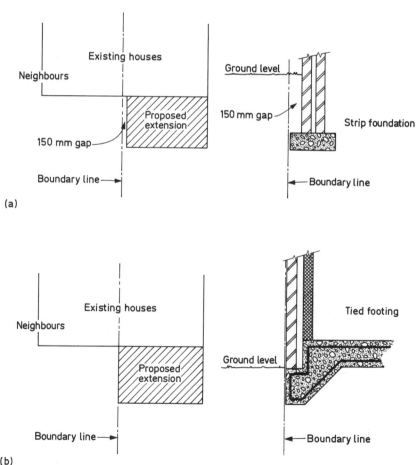

Fig. 8.13 The unco-operative next door neighbour: (a) first solution; (b) second solution.

as detailed in Fig. 8.13(b). As the toe of the tied footing does not need spread, the external wall face of the extension can be built on the boundary. Admittedly, the bricklayer building the wall adjacent to the neighbour may need to work 'overhand' to construct the wall on the boundary, but this does not normally prevent the extension being built. (Tied footings are dealt with in greater detail in later chapters.)

Privacy distances . . . parallel windows

As discussed before, most Local Authorities now have rules regarding 'overlooking'. Where there are two parallel blocks of houses (Fig. 8.14), most Planning officers will be concerned with maintaining reasonable privacy distances between windows of neighbouring houses. The reason is obvious. It would be very distasteful if one neighbour were allowed to construct a structure very close to, say, the main living room or bedroom window of another dwelling and thereby deprive the second party of his/her privacy. Every council has its own policies on such matters and you should enquire what they are.

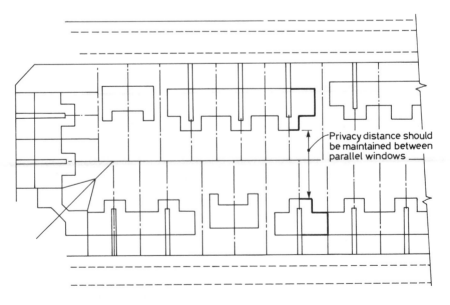

Privacy distance should be maintained between parallel windows

Fig. 8.14 Privacy distances

PART THREE: More on Building Control

9 More on building control

As explained in previous chapters, there are two Local Authority bodies that control simple development – the Planning Department and the Building Control Department. It is the latter that enforces the Building Regulations. (See below regarding Approved Inspectors.)

The Building Regulations are not uniform throughout the UK. (Scotland and Northern Ireland have their own regulations.) All references in this book are to the current Building Regulations which are in force in England and Wales. The Building Control system was revised in 1985 and new Building Regulation documents issued. Since then, there have been further amendments, the latest being the 1994 amendments (also known as the 1995 edition because they come into operation on 1st July 1995). It is essential that you obtain a copy of these regulations if you intend to become deeply involved with plan preparation. (Scottish or Northern Irish readers should obtain copies of their own regulations.)

These regulations are to control how a building is constructed and **not** what it looks like. As previously indicated, the Building Regulations were totally revised in 1985. Although the Regulations themselves now are shorter than the previous regulations, they now refer to subsidiary documents (called Approved Documents) and taken as a whole the Building Regulations are far longer and more comprehensive than ever before. Since 1985, various amendments have been introduced and the original Approved Documents have largely been replaced.

THE APPROVED DOCUMENTS

The Approved Documents quite clearly state that they are:-

> 'intended to provide guidance for some of the more common building situations. In other circumstances, alternative ways of demonstrating compliance with the requirements of the Building Regulations may be appropriate. There is no obligation to adopt any particular solution contained within the Approved Document . . . If you have not followed the guidance, it will then be for you to demonstrate by other means that you have satisfied the requirements.'

In other words, the solutions indicated in the Approved Documents are not compulsory, but the onus of proving that you have complied with the Building Regulations falls on the applicant/agent.

When dealing with simple structures such as domestic extensions/new dwellings, following the laid down guidelines is usually the sensible approach unless for some reason alternatives have to be found.

The list of Approved Documents in support of the Building Regulations in force at the time of writing is as follows:

- Reg 7 – Materials and Workmanship (1992 Edition)
- Approved Document A – Structure (with requirement A4 deleted 1994)
- Approved Document B – Fire Safety (1992 Edition)
- Approved Document C – Site Preparation and Resistance to Moisture (1992 Edition)
- Approved Document D – Toxic Substances (1985 Edition)
- Approved Document E – Resistance to Passage of Sound (1992 Edition)
- Approved Document F – Ventilation (1995 Edition)
- Approved Document G – Hygiene (1992 Edition)
- Approved Document H – Drainage and Waste Disposal (1990 Edition)
- Approved Document J – Heat Producing Appliances (1990 Edition)
- Approved Document K – Stairways, Ramps and Guards (1992 Edition)
- Approved Document L – Conservation of Fuel and Power (1995 Edition)
- Approved Document M – Access for Disabled People (1992 Edition)
- Approved Document N – Glazing materials and protection (1992 Edition)

The effects of these Approved Documents will be considered in greater detail later in Part Four. However, as these documents are constantly being updated, it is essential for anyone involved in the building industry to keep abreast of the changes.

THE PURPOSE OF THE BUILDING REGULATIONS/APPROVED DOCUMENTS

When plans are sent into the Building Control Department at the local council, the Building Control Officer (BCO) in charge of the application will check and ensure that the plans comply with the Regulations. For example, the BCO will check to see that the plans indicate that the floor joists and roof joists are the correct size and that an adequate damp-proof course has been shown built into brick-built external walls of habitable rooms. He/she will also ensure that the DPC is positioned correctly at minimum 150 mm (6 in) above external ground level. You can rest assured that the BCO will also work through a lengthy checklist to ensure that the plans are amended before any approvals are given. If the amendments are not made then the plans will be rejected.

As well as checking the plans, once work has commenced on site, it is usual for the Building Control Department to send a BCO out at regular intervals to ensure that the work complies with the approved plans, especially when the foundations are being constructed.

In order that the BCO knows when works are ready for inspection, there is a requirement that the builder/owner sends in stage notification cards at regular intervals. If these cards are not sent in and the work proceeds without the BCO being notified in due time, he/she can require the works to be opened up to ensure that the builder has complied with the Regulations. (**Note**: the BCO is not duty bound to make these inspections but the builder/owner has to let him/her know when the works are ready for inspection.)

Where appropriate, the BCO may also make observations or issue instructions to the builder should site conditions warrant it. (For example, if firm or unfilled ground cannot be reached at normal depths, it is sometimes necessary for the builder to excavate the foundation trenches deeper than originally envisaged. Or, if the original assumptions shown on the approved plans need revising, the BCO may offer suggestions regarding the drainage system.)

In my experience, because they are very busy people, it is unusual for the Planning Officer to visit the site of a domestic extension during the construction period (they do normally make a site visit prior to granting approvals). However, most BCOs do liaise with Planning as the work proceeds. (This is not always the case. I have been informed by the NHBC that surveys that they have carried out indicate that only 50% of Local Authorities have established arrangements for BCOs to check on planning conditions.)

However, if the Authority concerned does have such arrangements, if the terms of the planning approval are contravened, you can be reasonably certain that the BCO will pass the information on to his colleagues in Planning and that someone from that department will swiftly appear and insist that the works be carried out in an approved manner. (**Note**: As both the Planning and Building Control Departments have recourse to law if their decisions are not obeyed, it is unwise to ignore all reasonable instructions.)

APPROVED INSPECTORS

Legislation contained within The Building Act 1984 made it possible for the supervision of building control matters to be carried out either by Local Authority Building Control Officers or by Approved Inspectors. At the moment there only seems to be one body that has taken up the challenge, namely the National House-Building Council (NHBC). From recent discussions with the NHBC, I understand that they do not at this present moment provide inspection services in connection with small domestic extensions but will deal with such matters as the inspection of individual new houses, housing estates and major conversion work (e.g the conversion of an old barn into, say, flats.) The NHBC is now issuing its own standards for house building work, *NHBC Standards* (Volumes 1 and 2) from which I have quoted extensively and would recommend that you purchase a copy of these excellent publications which complement and amplify the Building Regulations.

FULL PLANS/BUILDING NOTICE

The 1985 regulations also brought in procedural changes. In the past, plans had to be submitted before work could commence on site. It is now possible to adopt two procedures which are as follows:

(a) Send a building notice to the Local Authority.
(b) Deposit full plans.

The idea behind the building notice system is that it will speed up building operations in as much that work can start on site virtually immediately and without having to have plans approved prior to commencement. Owing to the problems that could reasonably be anticipated with the first option – the building notice – I would not recommend this procedure except in very exceptional circumstances. You as the designer/builder could be found liable in negligence if work carried out on the site does not comply with Building Regulations.

The building notice system cannot be used for offices, shops, factories (i.e. designated buildings).

This book assumes that you will adopt the second option – the full plans application – as this system protects the various parties against contravention of the Building Regulations. If you need to know more about the building notice system, then it is usually possible to obtain the relevant information on how it works from your local Building Control Department.

10 Is building control approval required?

Not every extension to a domestic property requires building control approval because like the Planning Department, Building Control do not wish to be flooded by hundreds of very minor applications that would be unlikely to create any major problems.

The types of exempt work most likely to be encountered when dealing with domestic alterations and improvements are as follows:

(a) Porches.
(b) Conservatories (under 30 m^2 of floor area).
(c) Covered ways open on at least two sides (ditto).
(d) Car ports open on at least two sides (ditto).

Note the following:

(a) A conservatory or porch must satisfy the requirements of part N of the Building Regulations. Safety glass must be incorporated where required by the Regulations.
(b) Just because the Building Regulations do not apply to these structures under the new regulations does not mean that planning permission is not required (refer to previous chapters). As with Planning, even if you believe that something that you are designing is exempt under Building Regulations, I would advise you to send a copy of the plan to the Building Control Department and ask for confirmation that building control approval is not applicable. Be safe, not sorry.

If your proposals do not come within the exempt category (kitchen extensions, lounge extensions and the like are **not** exempt), then you will need to apply for building control approval, and filling out the forms is dealt with in the next chapter.

11 Making the building control submission

FILLING IN THE FORMS

In order to obtain building control approval using the 'full plans submission' system, your plans have to be submitted to the relevant Local Authority together with a suitable application form. As with the planning application described before, under normal circumstances, Local Authorities have their own specially printed application forms. Each Authority's forms vary slightly but the basic layout is the same.

I have reproduced a typical building control form and fee scales in Fig. 11.1 and 11.2. If you examine Fig. 11.1(a) and (b) you will note that a building control application form is very similar to a planning form and on the example shown requires answers to 11 questions.

Questions 1 and 2: question 1 asks for the name and address of the applicant. You should provide exactly the same information as you supplied on the planning form. The applicant is the person who wishes to have the works carried out (e.g. the home owner). Question 2 asks for the name of the agent. Once again, you should provide exactly the same information as you supplied on the planning form. If you are working for someone (you have prepared the plans for someone other than yourself) and are looking after the submission, then you are the agent and you must insert your name and address in order that all correspondence will be directed to you. If you do not give your name, any queries will be directed to the client. If you are not acting for someone (e.g. the application is for an extension to your own house), then there is no agent, so leave the box blank or cross it through. If you have a telephone number, provide it because if there are any queries, then someone can contact you easily and it helps to prevent delay.

Question 3 asks for the address applicable to the application. This address is not necessarily the same as the applicant's address. You or your client could be living at one address but own another property.

Question 4 asks for particulars of the development. Once again try to use the same description as you used on the planning application. Do not write down a massive rambling description.

The Building Act 1984
The Building Regulations 1991

FULL PLANS APPLICATION

This form is to be filled in by the person who intends to carry out building work or agent. If the form is unfamiliar please read the notes on the reverse side or consult the office below.
Please type or use block capitals.
Halton Borough Council, Municipal Building, Kingsway, Widnes, Cheshire. WA8 7QF Tel. 051 424 2061

HALTON
RUNCORN · WIDNES

Local Authority
BUILDING CONTROL

---

1 Applicants details (see note 1)

Name : *MR & MRS J SOAP*

Address : *7, SOMEWHERE ROAD, HALTON*

Postcode : *H1 XYZ*   Tel : ~~   ~~   Fax :

---

2 Agent's details (if applicable)

Name : *ANDREW R WILLIAMS & Co.*

Address : *437, WARRINGTON ROAD, RAINHILL.*

Postcode : *L35 4LL*   Tel : *0151 426 9666*   Fax : *0151 493 1425*

---

3 Location of building to which work relates

Address : *7, SOMEWHERE ROAD, HALTON*

Postcode : *H1 XYZ*   Tel : ~~   ~~   Fax :

---

4 Proposed work

Description : *DINING AND KITCHEN EXTENSION (SINGLE STOREY)*

---

5 Use of building

1   If new building or extension please state proposed use : *DWELLING 1(c)*

2   If existing building state present use : *AS ABOVE*

3   Is the building to be put, or intended to be put, to a use which is designated for the
   purpose of the Fire Precautions Act 1971 (see note 3)   ~~YES~~/NO

---

6 Condition approval (see note 4)

Do you consent to the plans being passed subject to conditions where appropriate ?   YES/~~NO~~

---

7 Fees (see note 5)

~~NEW~~ DWELLINGS. Number of Units   *N/A*   Number of house types

EXTENSIONS TO DWELLING. Total internal floor areas of extension(s) *UNDER 40*   m²

ANY OTHER WORKS (including extensions over 60m²).

Total estimated cost of work (excluding V.A.T.) : £ *N/A* . :

Fee :	£	*60*	*00*
Plus V.A.T. :	£	*10*	*50*
Total Fee Payable :	£	*70*	*50*

CONTINUED OVERLEAF

Fig. 11.1(a), (b) Full plans application. (Reproduced with permission of Halton Borough Council.)

8   Statement

This notice is given in relation to the building work as described, and is submitted in accordance with Regulation 11(1)(b) and is accompanied by the appropriate fee. I understand that **further** fees may be payable following the first inspection by the local authority.

Signature : ~~~~~   Name : *A N D R ew R . W i L L i A m s*   Date : — —

9   Extension in time :

There are many occasions when due to the need for additional information or amended plans to be submitted, it is not possible to give a favourable decision in the statutory period of 5 weeks. Do you agree to an extension of time of 3 weeks in situations such as this, please indicate and sign below.

YES/NO

Signed : ~~~~~ .   ~~Owner~~/Agent

10   Have you applied for Planning Permission   YES/NO

Note: This application relates only to Building Regulation procedures, you may also need approval under the Town and Country Planning Acts. Please contact Development Control. Tel. 051 424 2061

11   Resubmission

Has any previous application been submitted in respect of this scheme or project? Please give details including :   Date : N/A .   Ref. No. : —————

Notes

1. The applicant is the person on whose behalf the work is being carried out, e.g. the building's owner.

2. Two copies of this notice should be completed and submitted with plans and particulars in duplicate in accordance with the provisions of the Building Regulation 13.

Subject to certain exceptions where Part B (Fire Safety) imposes a requirement in relation to proposed building work, two further copies of plans which demonstrate compliance with the requirements should be deposited.

3. Premises currently designated for the purpose of the Fire Precautions Act 1971 are:

- Premises within the Fire Precautions (Hotels and Boarding Houses) Order 1972.
- Premises within the Fire Precautions (Factories, Offices, Shops and Railway Premises) Order 1989.

4.   Conditional Approval

Section 16 of the Building Act 1984 provides for the passing of plans subject to conditions. The conditions may specify modifications to the deposited plans and/or that further plans shall be deposited.

5. Subject to certain exceptions a Full Plans Submission attracts fees payable by the person by whom or on whose behalf the work is to be carried out. Most fees are payable

in two stages. The first fee must accompany the deposit of plans and the second fee is payable after the first site inspection of work in progress. This second fee is a single payment in respect of each individual building, to cover all site visits and consultsations which may be necessary until the work is satisfactorily completed.

The appropriate fee is dependent upon the type of work proposed. Fee scales and methods of calculations are set out in the Guidance Notes on Fees which is available on request.

6. Subject to certain provisions of the Public Health Act 1936 owners and occupiers of premises are entitled to have their private foul and surface water drains and sewers connected to the public sewers where available. Special arrangements apply to trade effluent discharge. Persons wishing to make such connections must give not less than 21 days notice to the appropriate authority.

7. Persons proposing to carry out building work or make a material change of use of a building are reminded that permission may be required under the Town and Country Planning Acts.

8. Further information and advice concerning the Building Regulations and planning matters may be obtained from your local authority.

Fig. 11.1(b)

Question 5 is aimed at establishing the use, something that might not be obvious just by looking at the plans. Approved Document B (more details will be given later), Table D1 classifies buildings into various purpose groups. Residential (dwellings) are divided into three parts:

1(a) Flats/maisonettes.
1(b) Dwelling house with floors above 4.5 m above ground level.
1(c) Ditto but with no floors above 4.5 m above ground level.

For most simple applications the group will therefore be a dwelling 1(c). The answer to 5/3 will be 'No' when dealing with a dwelling.

Question 6 allows the Building Control Department the authority to pass the plans with conditions. I usually answer 'Yes'. If you prevent the Building Control Officer from having the right to condition the approval, the officer might be forced to reject it because you have not allowed him/her to do otherwise.

Question 7 deals with fees. As with planning applications, fees are payable when an application is submitted. Figure 11.2(a) and (b) indicates the fee scales applicable to building control applications as at the time of writing. At present, unless the domestic extension is under 6 m^2 of floor area, the fees are paid in two parts. There is a plan submission fee (see plan fee column) and this fee has to be sent in with the forms (including the VAT). If you are carrying out work for a disabled person, fees are sometimes waived, but proof has to be provided to substantiate the disablement (e.g. a doctor's letter). Work that is obviously not entirely for the use of a disabled person will attract a fee. Resubmissions (see also question 11) do not normally require a fee to be sent with them unless the plans submitted are nothing like the original application.

As fees change from time to time, it will be necessary to obtain copies of the current plan submission fees. A second fee is payable once the work starts on site. If you are acting as the agent, this fee does not concern you if you are using the full plans system because you only require the plan fee at the time of submission. However, it is worth telling the client that he or she will eventually have to pay this sum, otherwise it may come as a nasty shock.

A 'tip' for agents/consultants/surveyors concerning fees: the current plan submission fees for domestic extensions can be very high (see Fig. 11.2(b)). I always ensure that my clients provide me with signed cheques for Local Authority fees. I normally ask for crossed cheques made out to the Local Authority. If the Local Authority don't agree with the fee they will notify you accordingly. Doing this avoids having to account for holding 'clients' money'. It also ensures that you are not in the position of having to pay out money in advance of being paid by the client, thus saving interest charges at the bank.

Question 8 is to ensure that the Local Authority are paid their second fee once work starts on site.

HALTON

BUILDING CONTROL

FEES FOR CERTAIN SMALL DOMESTIC BUILDINGS AND EXTENSIONS

Type of Work	PLAN FEE Fee £	V.A.T. £	Total £	INSPECTION FEE Fee £	V.A.T. £	Total £	BUILDING NOTICE FEE Fee £	V.A.T. £	Total £
1 Any extension of a dwelling (not falling within entry 4 below) the total floor area of which does not exceed 6m², including means of access and work in connection with that extension.	100.00	17.50	117.50	*PAID ON SUBMISSION*			100.00	17.50	117.50
2 Any extension of a dwelling (not falling within entry 4 below) the total floor area of which exceeds 6m², but does not exceed 40m², including means of access and work in connection with that extension.	60.00	10.50	70.50	150.00	26.25	176.25	210.00	36.75	246.75
3 Any extension of a dwelling (not falling within entry 4 below) the total floor area of which exceeds 40m² but does not exceed 60m², including means of access and work in connection with that extension.	85.00	14.88	99.88	220.00	38.50	258.50	305.00	53.38	358.38
4 Any extension or alteration of a dwelling consisting of the provision of one or more rooms in roof space, including means of access.	85.00	14.88	99.88	220.00	38.50	258.50	305.00	53.38	358.38
5 Erection of a detached building which consists of a garage or carport or both having a floor area not exceeding 40m² in total and intended to be used in common with an existing building, and which is not an exempt building.	17.00	2.98	19.98	53.00	9.28	62.28	70.00	12.25	82.25
6 Installation of cavity fill insulation in accordance with Part D of Schedule 1 to the Principal Regulations, where installation is not certified to an approved standard or is not installed by an approved installer, or is not part of a larger project.	*NOT APPLICABLE*			50.00	8.75	58.75	50.00	8.75	58.75
7 Installation of an unvented hot water system in accordance with Part G3 of Schedule 1 to the Principal Regulations, where the installation is not part of a larger project and where the authority carry out an inspection.	*NOT APPLICABLE*			50.00	8.75	58.75	50.00	8.75	58.75

FEES FOR NEW DWELLINGS

PLAN FEE

Number of dwellings (1)	Basic Fee (2) Fee £	V.A.T. £	Total £	Additional fee for each dwelling above the minimum number in the band in column (1) (3) Fee £	V.A.T. £	Total £
1	80.00	14.00	94.00	–	–	–
2-5	120.00	21.00	141.00	40.00	7.00	47.00
6-10	270.00	47.25	317.25	30.00	5.25	35.25
11-20	410.00	71.75	481.75	20.00	3.50	23.50
MORE THAN 20	800.00	105.00	705.00	10.00	1.75	11.75

INSPECTION FEE

Number of dwellings (1)	Further fee for each type of dwelling in excess of one (4) Fee £	V.A.T. £	Total £	Basic Fee (5) Fee £	V.A.T. £	Total £	Additional fee for each dwelling above the minimum number in the band in column (1) (6) Fee £	V.A.T. £	Total £
1	–	–	–	180.00	31.50	211.50	–	–	–
2-5	70.00	12.25	82.25	345.00	60.38	405.38	165.00	28.88	193.88
6-10	70.00	12.25	82.25	990.00	173.25	1,183.25	150.00	26.25	176.25
11-20	70.00	12.25	82.25	1715.00	300.13	2015.13	125.00	21.88	146.88
MORE THAN 20	70.00	12.25	82.25	2940.00	514.50	3454.50	100.00	17.50	117.50

Fig. 11.2(a), (b) Scale of fees (at time of writing). (Reproduced with permission of Halton Borough Council.)

FEE FOR DOMESTIC ALTERATIONS (AND EXTENSIONS OVER 60M²) AND ALL INDUSTRIAL AND COMMERCIAL WORKS

Total Estimated Cost	PLAN FEE			INSPECTION FEE			BUILDING NOTICE FEE		
	Fee £	V.A.T. £	Total £	Fee £	V.A.T. £	Total £	Fee £	V.A.T. £	Total £
1 Total estimated cost of work is £2,000 or less	60.00	10.50	70.50	PAID ON SUBMISSION			60.00	10.50	70.50
2 Total estimated cost of work exceeds £2,000 but does not exceed £5,000	150.00	26.25	176.25	PAID ON SUBMISSION			150.00	26.25	176.25
3 Total estimated cost of work exceeds £5,000 but does not exceed £20,000	37.50	6.56	44.06	112.50	19.69	132.19	150.00	26.25	176.25
plus for every £1,000 (or part) by which the cost exceeds £5,000 add	2.50	0.44	2.94	7.50	1.31	8.81	10.00	1.75	11.75
4 Total estimated cost of work exceeds £20,000 but does not exceed £100,000	75.00	13.13	88.13	225.00	39.38	264.38	300.00	52.50	352.50
plus for every £1,000 (or part) by which the cost exceeds £20,000	2.00	0.35	2.35	6.00	1.05	7.05	8.00	1.40	9.40
5 Total estimated cost of work exceeds £100,000 but does not exceed £1,000,000	235.00	41.12	276.12	705.00	123.38	828.38	940.00	164.50	1104.50
plus for every £1,000 (or part) by which the cost exceeds £100,000	1.25	0.22	1.47	3.75	0.66	4.41	5.00	0.88	5.88
6 Total estimated cost of work exceeds £1,000,000 but does not exceed £10,000,000	1360.00	238.00	1598.00	4080.00	714.00	4794.00	5440.00	952.00	6392.00
plus for every £1,000 (or part) by which the cost exceeds £1,000,000	0.88	0.15	1.03	2.62	0.46	3.08	3.50	0.61	4.11
7 Total estimated cost of work exceeds £10,000,000	9235.00	1816.13	10851.13	27705.00	4848.38	32553.38	36940.00	6464.50	43404.50
plus for every £1,000 (or part) by which the cost exceeds £10,000,000	0.75	0.13	0.88	2.25	0.40	2.65	3.00	0.53	3.53

EXAMPLE (FOR 3 TO 7 ABOVE)

say total estimated cost of your work is £7,250.00 - from 3 above

PLAN FEE for first £5,000 =   £37.50
for each extra part £1,000 over £5,000, multiply by £2.50 i.e. 3 x £2.50 =   £7.50
so total **PLAN FEE** =   £45.00 + 17½ % V.A.T.

INSPECTION FEE for first £5,000 =   £112.50
for each extra part £1,000 over £5,000, multiply by £7.50 i.e. 3 x £7.50 =   £22.50
so total **INSPECTION FEE** =   £135.00 + 17½ % VAT

The **BUILDING NOTICE FEE** is the **PLAN FEE** plus the **INSPECTION FEE** added together and is paid **on submission**

Fig. 11.2(b)

Question 9 By signing this section, you agree to give the Local Authority additional time to consider the application. I always mark this 'Yes'. If you refuse to give the Building Control Officer additional time and the plans are not quite right then he/she will be forced by law to reject the application as it cannot be amended within the statutory five-week period.

Question 10 asks if you have applied for planning permission. It is designed to ensure that planning procedures are not forgotten.

Question 11 covers resubmitted applications (i.e. an application that has been refused on technical grounds but has been amended to comply with the Building Regulations). It is necessary to give the date and reference number of the original application so that the Building Control Department can cross-reference with the original.

SENDING THE DOCUMENTS IN

When submitting your documentation read the forms carefully. Irrespective of what may be laid down by the Regulations, most Authorities have their own requirements regarding numbers of forms and plans that they require. The forms issued by your Building Control Department usually stipulate the numbers of copies of each document that they need. If you do not provide the correct number, you could delay your client's application.

Do not forget to send a cheque for the relevant plan fee. The Authority will not accept the application unless the fees are included.

QUERIES RAISED BY BUILDING CONTROL

Unlike the Planning Department, who rarely contact the applicant/agent once the plans are lodged (unless there is a serious non-conformity with established policy) it is virtually certain that you will receive a notification from the Building Control Department informing you that your plans are defective and do not comply with the Building Regulations.

When you receive a notification of this nature, there is a tendency for the inexperienced agent or householder to be either insulted or very concerned. As I said before, queries are the norm because the Building Regulations are very complex, and even someone who has a very good working knowledge of the Building Regulations can omit to include vital information. If a plan passes without any queries then you can count yourself as being very lucky.

Sometimes if amendments are only minor a letter may be issued requesting authorization to amend the plans.

ACTION TO BE TAKEN UPON RECEIPT OF A NOTICE OF NON-COMPLIANCE

I would advise that you take the following action:

(a) Telephone the Building Control Officer and clarify any points which are not obvious. Most forms tend to be couched with vague references to the Building Regulation clauses. Find out what is not considered correct and try to rectify the omission. Sometimes your plan may actually cover the point in question and the officer has missed it because he or she has not had the time to fully check the application (everyone is human and errors can occur).

(b) Fill in the extension of time form and post it back (if you have not already granted an extension on the form).

(c) Amend the master copy of your plans immediately and send revised sets of copy plans to the Building Control Officer with a covering letter stating clearly the application number that has been issued to your submission.

RECEIVING A REJECTION NOTICE

There will be occasions when your plans will just be stamped rejected and sent back. When this happens, once again, there is a strong temptation for the inexperienced home owner/agent to dress up in sackcloth and ashes and bewail his/her lot, especially if no notice informing you of non-compliance has been received.

Although Building Control Officers very rarely admit to being overloaded with work, they are sometimes forced to 'defend their corner' by clearing out those applications which either cannot be dealt with in the time available or which require a large number of amendments making.

A rejection notice does not automatically mean that the Building Control Officer considered the plan sent in so terrible that it was beyond redemption. It should be borne in mind that Building Control are obliged to pass or reject an application within a set timescale (five weeks, or eight weeks if an extension of time has been granted).

ACTION TO BE TAKEN IF PLANS ARE REJECTED BY BUILDING CONTROL

I would advise that you take the following action:

(a) Once again, telephone the Building Control Officer and clarify any points which are not obvious. (Do not be rude or offensive. The person on the other end of the 'phone is only doing his/her job.) Find out what is considered incorrect and try to rectify the omission/mistake.

(b) Resubmit the application as quickly as possible after making any amendments, but make sure that you indicate that the application is a resubmission and that no further fee is payable. (This is where question 11 comes in.)

PART FOUR: Building Construction

12 Notes concerning new Approved Document L – Conservation of Fuel and Power (1995 Edition)

Because of the fears concerning global warming, Approval Document L (Conservation of Fuel and Power) has now been completely revised, and as from 1st July 1995, new thermal standards and a SAP rating system will be introduced. SAP stands for Government Standard Assessment Procedure, and under the new regulations, dwellings have been divided into two groupings – those with SAP ratings under 60 and those with SAP ratings over 60 (refer to page 8 of Approved Document L). As from 1st July 1995 all **new** dwellings will have to be given a SAP rating when the plans are sent into Building Control for checking.

I have to confess, when I first studied the new regulations, they filled me with horror. Was the new SAP system going to be applied to house extensions? If it was, did SAP calculations have to be prepared for the old part of the house or not? After referring to Table 1 (page 8 of Part 1), it appeared on first sight as if they did.

After carrying out one or two trial SAP calculations on typical older houses, I have come to the conclusion that a large number (millions) of existing houses (those built prior to the introduction of thermal regulations) could have ratings as low as 11. But this is hardly surprising, because some properties still have little or no loft insulation. Many houses have walls that are built of brick, non-insulating block or possibly no-fines concrete. Very few have any form of floor insulation.

So, did a person constructing an extension have to super-insulate the extension if the existing house was badly insulated or not? After searching through the new regulations, I found Clause 0.3 on page 6. This clause states that 'Where an extension does not exceed 10 m^2 of floor area that reasonable provision may be considered to have been made if the construction is no less effective for the purposes of the conservation of fuel and power than the existing construction'. Although this clause is obviously intended to ensure that designers and builders of small extensions were not hit with the full force of the new regulations, it is vague in the extreme. For instance, how does one prove what the existing structure consisted of to start with? And what if the extension exceeds 10 m^2 of floor area?

However, an article in the *Institute of Building Control News* (October 1994) indicates that SAP rating will not be applied to house extensions. According to their report, Tony Fields of the Department of the Environment explained that SAP energy ratings only apply to new dwellings and that the insulation of domestic extensions will continue to be based on the elemental approach using *U*-values (the coefficient of thermal transmittance). Furthermore, it is considered appropriate to adopt the *U*-values relating to the higher SAP rating shown in Diagram 2 (page 9 of Approved Document L), reproduced below.

Element of structure	*U*-value (W/m^{2}K)	
Roofs	0.25	(0.35 flat roofs[a])
Exposed walls	0.45	
Exposed floors and ground floors	0.45	
Semi-exposed walls and floors	0.60	
Windows, doors and rooflights	3.30	

[a]See note 2 on Table 1 (page 8 of Approved Document L) which indicates that 0.35 W/m^{2}K will be acceptable for certain roofs.

The specifications given in this book are therefore designed around the above.

In Appendix B, I have reproduced my standard specification. If you refer to the various sections within the specification, you will note that I do not just quote the *U*-values given but supply the Building Control Officer (and the contractor) with details of the type of insulation and surrounding construction to be used to achieve the values given. Obviously when dealing with manufactured products, I have tried to specify manufactured sizes rather than the theoretical thicknesses that are deemed to comply with the various schedules within Approved Document L.

BUILDING CONSTRUCTION TERMINOLOGY

Like any other industry, the building industry has developed its own peculiar terminology for components that form a finished structure. These terms have grown up over the centuries and are the ones used in all technical documentation, including the Building Regulations. If you refer to Figs 12.1 and 12.2 some of the major elements are indicated. Some of my readership may be familiar with most of the terms but obviously, if you are not, then you

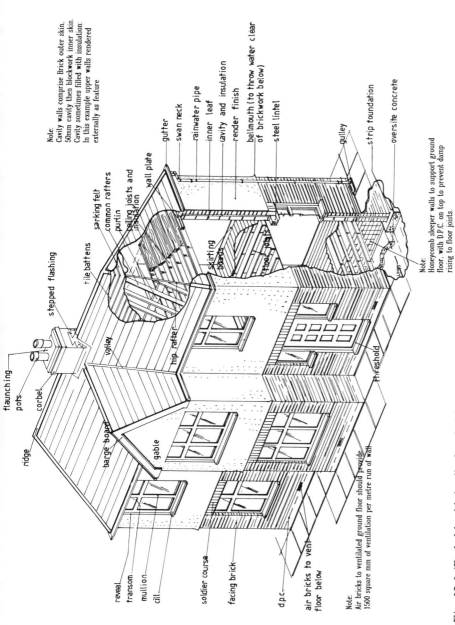

Note:
Cavity walls comprise Brick outer skin,
50mm cavity then blockwork inner skin.
Cavity sometimes filled with insulation.
In this example upper walls rendered
externally as feature

gutter

swan neck

rainwater pipe

inner leaf

cavity and insulation

render finish

bellmouth (to throw water clear
of brickwork below)

steel lintel

gulley

strip foundation

oversite concrete

Note:
Honeycomb sleeper walls to support ground
floor, with D.P.C. on top to prevent damp
rising to floor joists.

wall plate

sarking felt

common rafters

purlin

ceiling joists and
insulation

tile battens

stepped flashing

flaunching

pots

corbel

ridge

barge board

valley

hip rafter

gable

skirting
board

floor joists

threshold

reveal

transom

mullion

cill

soldier course

facing brick

d.p.c.

air bricks to vent
floor below

Note:
Air bricks to ventilated ground floor should provide
1500 square mm of ventilation per metre run of wall

Fig. 12.1 'Peeled back' detail of typical house.

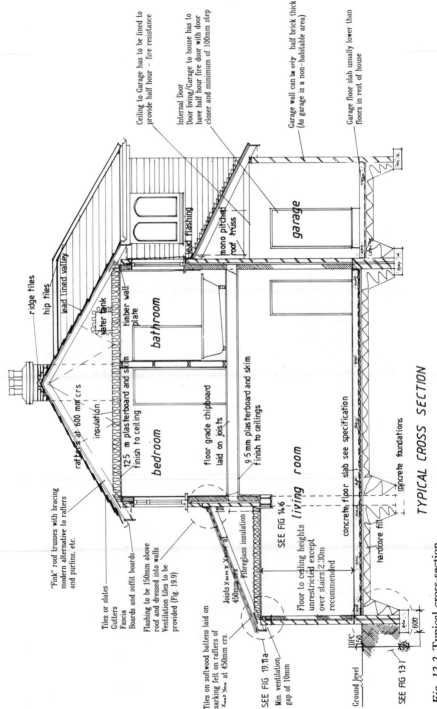

Fig. 12.2 Typical cross-section.

TYPICAL CROSS SECTION

Ceiling to Garage has to be lined to provide half hour – fire resistance

Internal Door
Door living/Garage to house has to have half hour fire door with door closer and minimum of 100mm step

Garage wall can be only half brick thick (As garage is a non-habitable area)

Garage floor slab usually lower than floors in rest of house

garage

mono pitched roof truss

lead flashing

ridge tiles

hip tiles

lead lined valley

water tank

timber wall plate

bathroom

rafters at 600 mm crs

insulation

125 m plasterboard and skim finish to ceiling

bedroom

floor grade chipboard laid on joists

9.5 mm plasterboard and skim finish to ceilings

living room

concrete floor slab see specification

concrete fundations

"Fink" roof trusses with bracing modern alternative to rafters and purlins. etc.

Tiles or slates
Gutters
Fascia
Boards and soffit boards

Flashing to be 150mm above roof and dressed into walls Ventilation tiles to be provided (Fig. 19.9)

Tiles on softwood battens laid on sarking felt on rafters of X mm x X mm at 450mm crs.

SEE FIG 19.1a
Min. ventilation gap of 10mm

Joists X mm x X mm at 450mm crs

Fibreglass insulation

SEE FIG 14.6

Floor to ceiling heights unrestricted except over stairs 2.30m recommended

hardcore fill

Ground level

DPC 150

SEE FIG 13.1

600

should examine the details carefully in order that you can understand the basics. As you will note, Fig. 12.1 shows a typical house with parts of the elements 'peeled back' in order to be able to view the underlying detail. Figure 12.2 is a cross-section of a typical building.

Obviously, these small scale details cannot convey the full picture. More detailed sketches have been provided later in the book when dealing with major elements (e.g. foundations, roofs etc.). When preparing a plan, it is essential that you use standard building terminology and are not tempted to invent your own. I have seen plans prepared by those who are not totally conversant with building constructing describing say, a purlin, as say, a roofbeam. This can be very confusing to builders and Building Control Officers.

For those unfamiliar with any building terminology, I have listed some of the major elements in a building and an outline description in Appendix C.

However, the list is not exhaustive and I would recommend that you consider buying a copy of the *Dictionary of Building* by John S. Scott (Penguin) or a similar publication.

13 The foundations

GENERALLY

The foundations of the house are probably one of the most important parts of the property because without an adequate base the property will quickly become unstable. But the foundations are only as strong as the substrata of ground upon which they are built.

Therefore, the substrata should:

(a) Be strong enough to sustain the loads put on it by the building.
(b) Not contain sulphates or other deleterious matter likely to destroy the foundations. (Alternatively, the concrete in the foundations will be of a mix capable of resisting the effects.)
(c) Not be susceptible to the effects of frost action. (Clay can expand if affected by frost. That is why gardeners like to ridge up clay soil in cold weather so it breaks up. If the subsoil is of a type susceptible to frost action then the underside of the concrete foundations must be built sufficiently deep that the soil beneath them will not freeze.)

However, foundation design can present a problem for anyone designing small domestic extensions because, unlike larger projects, you, the designer, are very much on your own. There are no other consultants such as engineers on hand to make soil tests and to advise you about ground conditions in the area.

On page 32 of Approved Document A (of the Building Regulations) the subsoils upon which your foundations are likely to be built are classified into seven types (e.g. rock, clay, sand) and there is also a list of field tests that you can carry out. For instance, silt (Class VI) is defined as 'fairly easily moulded in the fingers and readily excavated'. Fine, except that even a lay person knows that foundations are built below ground level. So, what is the point in attempting to assess the subsoil condition from ground level?

You could of course, visit your client armed with a pickaxe and spade and start by digging trial holes in their garden (trial holes are pits about 3 ft square excavated as deep as necessary to hit firm ground, and are sometimes excavated on larger sites in order to test soil conditions), but in my experience, most home owners do not want their builder/surveyor digging up their gardens, months in advance of work starting on site, especially as they know that they will have enough disturbance once work actually starts.

In any case, even if you did dig trial holes, it does not necessarily mean that they will give the full picture. It is not unknown for the original builders to construct houses on sites that are far from perfect. Whilst doing so, they often disguise problem ground and it is only when a new extension is being built that the defect is exposed.

There was a typical case in my office a few years back, when a new extension was in the process of being built. Everything went according to plan until the second builder began digging in the rear garden area and discovered the remains of an old pit. The first builder had decided that if he filled the pit with demolition rubble and old timber it would save him money. By doing so, he created a problem for the second builder because not only did he have to dig the new footings deeper than normal (normal strip footings cannot be built on filled ground), he was forced to clear out the buried debris and remove it, which costs money.

The fact is, when dealing with a small domestic extension, the best that you can do is make an assessment of the likely ground conditions, design for those conditions, and advise your client that the foundations by their very nature are subject to review once construction work starts on site.

However, you do not have to make your assessment totally blind. Without a doubt, your client will have done some gardening and if he/she has dug down to any depth will know what type of soil is there. Other people in the street may have had extensions built and may tell you what the ground was like. If you know any local builders or the local Building Control Officer they may also be able to advise you. So, ask the following questions:

(a) What is the soil like – is it clay, sand, rock?
(b) Did the neighbours have any problems when they built their extension and how deep did they need to built their foundations?

Past experience will also guide you. I know that on my 'patch', unless information supplied by my client or observations on site lead me to think otherwise, the ground is more than likely to be clay. (There is a brickworks not far away.)

FOUNDATION TYPES

There are several types of modern foundations for domestic construction which are:

(1) Traditional strip.
(2) Deep strip.
(3) Tied footing or raft.
(4) Piled (very unlikely to be used on a small extension). If piles are needed then specialist advice will be required anyway.

TRADITIONAL STRIP FOUNDATIONS

A typical traditional strip footing is indicated in Figs 13.1 and 13.2. Study the sketches carefully because you will have to use similar details on your drawings (albeit at a smaller scale).

Using the Approved Documents

As indicated previously, the Approved Documents quite clearly state that they are 'intended to provide guidance for some of the more common building situations . . . If you have not followed the guidance, it will then be for you to demonstrate by other means that you have satisfied the requirements'. In other words, the solutions indicated in the Approved Documents are not compulsory, but the onus of proving that you have complied with the Building Regulations falls on the applicant/agent. When dealing with simple structures such as domestic extensions/new dwellings, following the laid down guidelines is usually the sensible approach, unless for some reason alternatives have to be found.

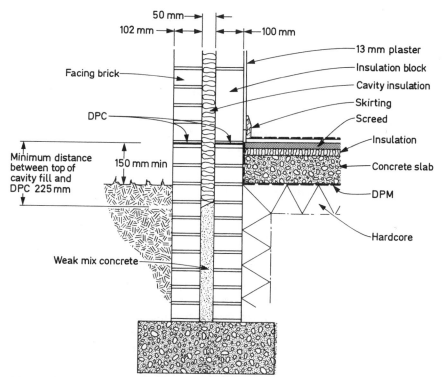

Fig. 13.1 Traditional strip footing: solid floors.

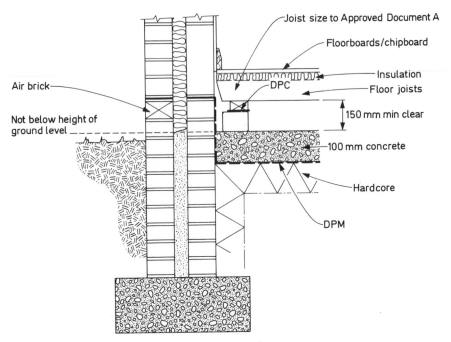

Fig. 13.2 Traditional strip footing: suspended floors.

Approved Document A, pages 31 and 32, states that 'Building Control' designed strip foundations of plain concrete should comply with the following rules:

(1) They should be built on firm ground. (If the soil is any type of backfill, then another type of foundation will be needed, unless it is possible to dig through the backfill and reach suitable ground.)

(2) The width of the concrete base must comply with Table 12 of Approved Document A (page 32). Table 12 grades the ground upon which buildings can be constructed into seven types (e.g. rock, clay) and the minimum strip foundation widths to use according to the calculated load per metre.

(3) The concrete used must be 1:3:6 mix (i.e. of a minimum consistency of 50 kg cement to not more than 0.1 m^3 of fine aggregate and 0.2 m^3 of coarse aggregate) or better, or Grade ST1. The minimum thickness of a foundation must be 150 mm (6 in) and that the thickness of a foundation must be not less than the projection. So if the concrete projects 200 mm for some reason, the thickness must be 200 mm. Where there are changes in level (stepped footing) the 'stepped' section must have minimum thickness of 300 mm. If piers are indicated on walls then the

footing must be enlarged to give the same projection from its face as the footing is from the main wall.

If you refer to my standard specification (Appendix B, clause 7), you will note that it indicates that foundation thicknesses shall be 200 mm. The reason for this is that if you take the wall thicknesses that we specify away from the width of the concrete footing, the projection exceeds 150 mm. The thickness of the concrete has therefore to be greater than 150 mm, and 200 mm is the next convenient thickness.

(4) In chemically aggressive soils, special concrete must be used. Sometimes it is necessary for the builder to use sulphate-resisting concrete. Sulphates in the ground (whether natural or manufactured) attack the concrete and, if present will eventually cause the foundations to disintegrate. As sulphate-resisting concrete is more expensive than concrete made with ordinary Portland cement, I am not suggesting that you specify it in all circumstances, but I am recommending that you put on your plan 'sulphate-resisting concrete if necessary'. This brings to the builder's attention to the fact that they must check the ground conditions before building. (It also covers the designer against claims from a possible irate client if the builder neglects to carry out the necessary checks.)

(5) Walls must be built centrally on a concrete footing when designing to Building Control standard specifications. The Building Control Officer will reject your plans if walls are built 'off centre' unless you can prove that the non-traditional foundation is acceptable.

Depths of foundations

If you refer back to Chapter 8 (Figs 8.11 and 8.12) you will note that it is essential when considering foundation depths to take into account any trees growing in the area. Even though the tree(s) in question might not be in the garden of the property in question, it (they) might have an effect on the proposed design.

On highly shrinkable clay soil, the NHBC set a minimum depth of 1 m as long as there are no trees present or recently removed from the area. Shallower foundations are allowed under NHBC regulations where the soil has a medium or low shrinkage potential.

However, I have had it brought to my attention that in some parts of the country, because of a spate of subsidence problems, that some local authorities are requiring foundation depths much deeper than 1 m.

On my 'patch', one Local Authority Building Control Department normally encourages designers not to dimension foundations at all but merely state on the plans a clause along these lines of 'the foundations shall be of a size and taken to a depth as approved by the Local Authority Building Control Officers'. Unfortunately, clauses of that nature, whilst giving the

Building Control Officer total control of that part of the works, do not help a tradesman builder who is trying to work out a quote for carrying out the works.

As it is unusual for a client to allow trial holes to be dug, I normally indicate on my plans that trial holes have not been taken and that the builder is to test as necessary. I indicate a minimum width of 600 mm on my plans, as this size is one of the approved dimensions on Table 12 of Approved Document A and allows for most ground conditions. It is also a convenient size for a bricklayer to work off.

DEEP STRIP/TRENCH FILL FOUNDATIONS

These are similar to traditional strip footings and must comply with the same rules, but the concrete is deeper as shown in Fig. 13.3. When market conditions are favourable and when ready-mixed concrete is cheap to buy,

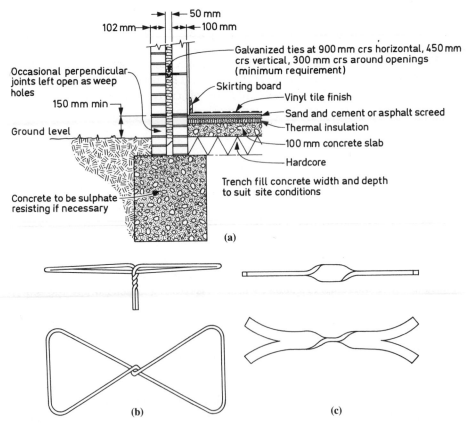

Fig. 13.3(a) Trench fill foundations. (b) Butterfly wall tie. (c) Vertical twist wall tie.

many builders prefer to use trench fill footings because by doing so, the top of the concrete is brought near to the surface of the ground and this cuts out bricklaying time in the foundations, which speeds up the job.

The NHBC requires that the minimum widths of any trench fill footing shall be at least 500 mm wide. Deep strip/trench fill footings are also used where ground conditions dictate deep excavation or where the ground is influenced by trees.

THE TIED FOOTING (raft foundation)

A tied footing is a very useful form of foundation but does not come within the scope of the Building Regulations. The simple form of tied footing comprises a reinforced floor slab with a toe beam (Fig. 13.4) at the edge. Because it is outside normal Building Control parameters, calculations are required by most Building Control Departments before tied footings can be used (Fig. 13.5(a) and (b) shows a typical set of calculations). If you study the calculation sheet you will note the engineer has proven that the steelwork in the slab and toe counteracts the overturning moment caused by non-central loading. It is a useful form of construction where it is necessary to build up to a boundary line but the next door neighbour will not allow the spread of a normal footing to pass under their garden. (See Chapter 8 for details.) I would suggest that it is always prudent to ask an engineer to prepare tied footing calculations unless you are very confident of your engineering knowledge. You will note from details given later on in the book that my standard conditions exclude the cost of engineer's fees from my fee.

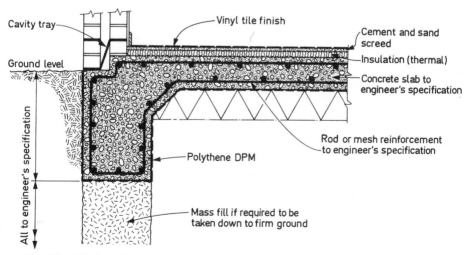

Fig. 13.4 Typical tied footing (raft foundation).

Contract			Job ref.
Part of structure Raft Foundation			Calc sheet No. 1
Drawing ref.	Calculations by	Checked by	Date

Members Ref.	CALCULATIONS	OUTPUT

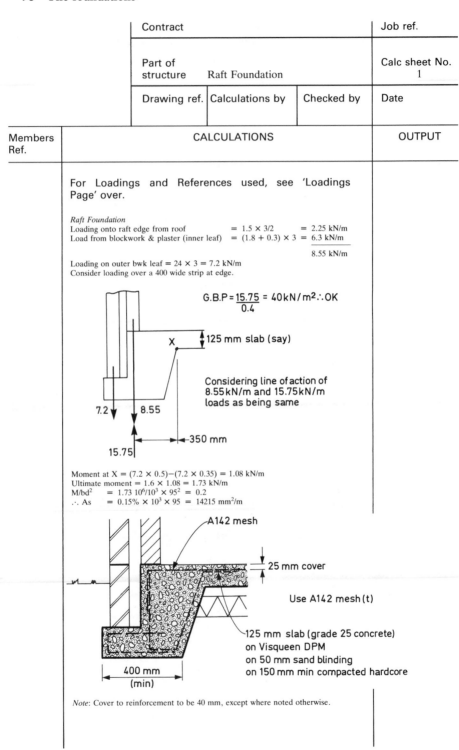

For Loadings and References used, see 'Loadings Page' over.

Raft Foundation
Loading onto raft edge from roof $= 1.5 \times 3/2$ $= 2.25$ kN/m
Load from blockwork & plaster (inner leaf) $= (1.8 + 0.3) \times 3 = 6.3$ kN/m
                                                                    8.55 kN/m

Loading on outer bwk leaf $= 24 \times 3 = 7.2$ kN/m
Consider loading over a 400 wide strip at edge.

G.B.P $= \dfrac{15.75}{0.4} = 40$ kN/m^2 ∴ OK

X   125 mm slab (say)

Considering line of action of
8.55 kN/m and 15.75 kN/m
loads as being same

7.2   8.55

350 mm

15.75

Moment at X $= (7.2 \times 0.5) - (7.2 \times 0.35) = 1.08$ kN/m
Ultimate moment $= 1.6 \times 1.08 = 1.73$ kN/m
M/bd^2 $= 1.73 \ 10^6/10^3 \times 95^2 = 0.2$
∴ As $= 0.15\% \times 10^3 \times 95 = 14215$ mm^2/m

A142 mesh

25 mm cover

Use A142 mesh (t)

125 mm slab (grade 25 concrete)
on Visqueen DPM
on 50 mm sand blinding
on 150 mm min compacted hardcore

400 mm
(min)

Note: Cover to reinforcement to be 40 mm, except where noted otherwise.

Fig. 13.5(a), (b) Raft foundation calculations.

Contract			Job ref.
Part of structure Loadings			Calc sheet No. 2
Drawing ref.	Calculations by	Checked by	Date

Members Ref.	CALCULATIONS	OUTPUT
	References	
	British Standard CP3: Chapter V: Part 1: 1967. British Standard BS 648: 1964.	
	Pitched Roof Loading	*kg/sq m*
	Interlocking tiles	50
	Battens and counterbattens	7
	Sarking felt	3
	Roof rafters say	13
	Fibreglass insulation	2
	Plasterboard and skim	20
	Superimposed loading	75
		———
		170
	Design loading taken as 1.7 kN/m²	
	Dormer Roof Loading	
	Chippings	20
	Three layers of roofing felt	6
	20 mm chipboard	15
	Roof joists and firrings say	10
	Fibreglass insulation	2
	Plasterboard and skim	20
	Superimposed loading	75
		———
		148
	Design loading taken a 1.5 kN/m²	
	Floor Loading	
	20 mm flooring grade chipboard	15
	Floor joists say	15
	Superimposed loading	150
		———
		180
	Design loading taken as 1.8 kN/m²	
	Partitions	
	Studding and fibreglass insulation say	7
	Plasterboard and skim	20
		———
		27
	Design loading taken as 0.3 kN/m²	

Fig. 13.5(b)

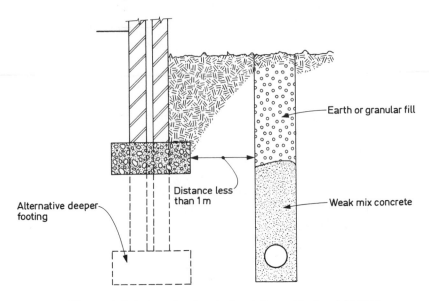

Where a drain is closer than 1 m of a foundation and the trench is lower
than the wall foundation then the trench must be concrete filled up to
the underside of foundation

(a)

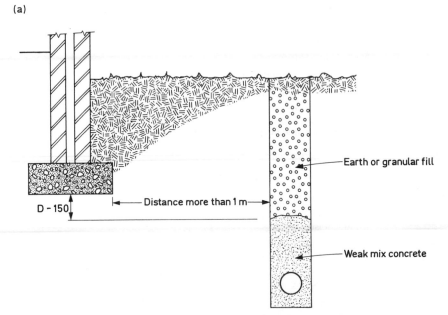

Where the foundation is over 1 m away and the drain is deeper than the
foundation then the concrete fill shall be not less than D – 150 mm

(b)

Fig. 13.6 Foundations near drains: (a) less than 1 mm away; (b) over 1 m away.

Experience has taught me that this is a wise precaution because whilst it is relatively easy to assess one's own work load, the likely engineering charge is unknown.

OTHER CONSIDERATIONS

Drains running near or through foundations

Sometimes existing drains on site can be deeper than the bottom of the proposed footing. The necessary measures to prevent collapse are indicated in Fig. 13.6(a) and (b). Where drain-pipes have to pass through foundation walls they should pass through a properly formed hole with a lintel over it so that they are not cracked if the building settles slightly (Fig. 13.7).

Suggested foundation note for drawing

Refer to clause 7 of Appendix B for standard foundations specification.

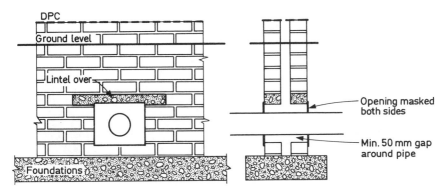

Fig. 13.7 Pipes passing through substructure walls.

14 External walls

BRICK BONDING

All brick and block walls should be built to a recognized bond. On external walls built in facing brick or selected commons the evenness of the bond is essential for appearances sake. (On walls that have a rendered finish, the neatness of the bonding is not so essential but the basics of bonding must be adhered to for stability.)

EXTERNAL WALLS GENERALLY

Nowadays, most external walls to modern domestic properties built in England and Wales are of cavity wall construction. Although there had been experimentation with cavity construction during the nineteenth century, cavity walls became more common at the beginning of the twentieth century in an attempt to overcome one of the major problems being experienced with solid walls, namely damp penetration.

Because stretcher bond is simple to use with cavity walls, it is the form of bonding most commonly seen on the external walls of modern housing (Fig. 8.3). There are other bonds, however (e.g. English bond (Fig. 8.4), English garden wall bond (Fig. 8.5) and Flemish bond (Fig. 8.6)). When you are carrying out your survey prior to preparing plans, note the brick bonding on the existing building. The bonds shown usually give a strong indication of the existing wall construction. Stretcher bond usually indicates that the walls are of cavity construction (but not always). English bond, Flemish bond and English garden wall bonds usually indicate solid wall construction. If you refer back to Chapter 8 ('Matching Materials') the subject of bond is discussed in greater detail.

Solid walls can still be used in modern construction, but now a water repellent defence has to be installed (e.g. external tile hanging or external rendering). As the cavity wall is now the most common, I intend to concentrate on this form of construction.

A cavity wall comprises two distinct skins of walling tied together with wall ties (Fig. 13.3(a) and (b) shows typical cavity wall tie details). The theory of the cavity wall is that the outer skin will obviously become wet when it rains but if the water manages to saturate the outer skin and forces its way inwards, when it reaches the cavity, it will merely run down the inner face of the outer skin and not dampen the inner leaf. Sometimes, even this defensive barrier is breached when building operatives become lazy and do not maintain a good standard of workmanship (e.g. bricklayers not keeping the cavity clear of mortar droppings). However, despite the human factor, there can be no doubt that cavity walls are far superior to solid walling in terms of weather resistance.

Approved Document L (see Chapter 12 for details) of the Building Regulations covers the standards required in modern buildings to help conserve fuel and power. As traditional building materials cannot easily provide the sort of thermal values needed to save power, it has become standard practice to incorporate insulating materials into external walls. This can be done by using insulating blockwork in one or both other skins of the hollow wall or by incorporating other insulants such as glassfibre in the walls. Alternatively, it is possible to use a combination of products.

In some instances the insulation is placed on the outside or on the inside of the external walls. However, doing this has its disadvantages (insulation tends to be thick and takes up valuable space.) Although it was resisted for a long time, it was inevitable that the void in the cavity wall should be targeted as a convenient area where the insulant could be sited without causing the external wall to become excessively thick.

Figure 14.1 shows typical cavity walls (both total fill and partial fill). As the name implies, the total fill cavity wall has all the cavity filled with an approved insulant such as Owens Corning's 'Dritherm' which is impregnated to resist water penetration. Naturally, filling the cavity brought back the dangers of damp penetration again and in order to ensure that damp problems experienced in solid walls do not re-occur, cavity filling can only be carried out if the fill material comes up to an approved standard.

The alternative to total fill is partial fill. In the standard specification (Appendix B, clause 12) a partial fill cavity system is indicated. Whether you use the total fill or partial fill system is a matter of designer's choice, but I would advise you to read pages 22 and 23 of *Thermal Insulation: Avoiding Risks* (HMSO) before specifying total fill. In exposure zones 1 and 2 (pages 1 and 2) total fill in walls without an impervious cladding or rendered finish will probably be acceptable. (**Note:** Scottish, Northern Irish and Manx readers should check their own Building Regulations on this point.)

Timber frame construction

There are other forms of external wall construction such as timber frame construction where the house is made in a factory and the brick outer skin is non-loadbearing. Although timber frame construction is beyond the scope of this book, it is worth a mention because if you own a timber framed house

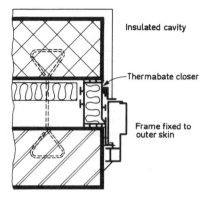

Insulated cavity

Thermabate closer

Frame fixed to
outer skin

Plan view showing cavity closings
using thermabate closers

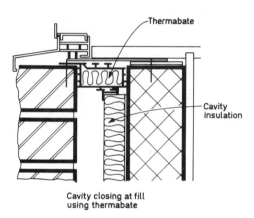

Thermabate

Cavity
insulation

Cavity closing at fill
using thermabate

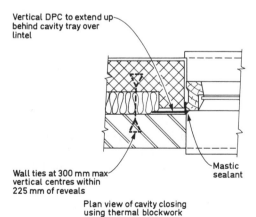

Vertical DPC to extend up
behind cavity tray over
lintel

Wall ties at 300 mm max
vertical centres within
225 mm of reveals

Mastic
sealant

Plan view of cavity closing
using thermal blockwork

Fig. 14.1 Cavity wall insulation details. (Details after Owens Corning Insulation Ltd.
and RMC Panel Products Ltd.)

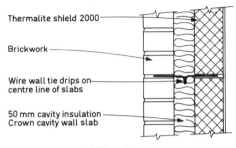

Thermalite shield 2000

Brickwork

Wire wall tie drips on centre line of slabs

50 mm cavity insulation Crown cavity wall slab

Total fill cavity wall

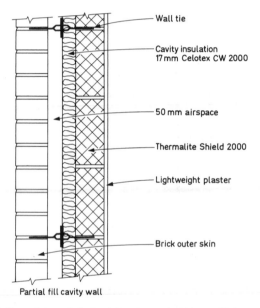

Wall tie

Cavity insulation 17 mm Celotex CW 2000

50 mm airspace

Thermalite Shield 2000

Lightweight plaster

Brick outer skin

Partial fill cavity wall

NB. Thermabate or similar insulating closers should be used if blockwork inner skin and blockwork 'closer' would create cold bridges

Fig. 14.1 contd.

and want to alter it or are asked to design plans for alterations to a timber framed house, beware. In my experience, expensive specialist calculations are sometimes needed before any alterations can be made to the framework.

CAVITY WALLS BELOW GROUND LEVEL

The walls below ground level usually comprise two half-brick skins, two 100 mm blockwork skins or one half-brick outer skin with a blockwork inner skin with a weak concrete cavity fill between them (Figs. 13.1 to 13.3). The purpose of the weak mix concrete cavity fill is to ensure that building operatives do not squash together the skins of the hollow wall below ground level as they backfill the trenches and destroy the stability of the foundations.

The weak concrete cavity fill has no other function and must be kept down from the damp-proof course (DPC) in the external wall. I usually indicate that the fill should be kept down 225 mm below the DPC level. In my opinion, if the weak concrete cavity fill is shown higher than this, there is a danger that any water collecting at the bottom of the cavity during wet weather might breach the DPC defence in the wall, especially if the bricklayers have allowed 'snots' of mortar to fall down the cavities when they are building the superstructure.

CAVITY WALLS ABOVE DPC LEVEL

As indicated above, in recent years, the demand for better thermal insulation standards has resulted in major changes. At the time of writing, external superstructure walls to habitable areas of a house should now have a U-value of 0.45 W/m^{2}K. (The U-value is the thermal transmittance coefficient).

Where cavity fill is used to attain the required U-value it has to comply with the requirements of Approved Document D of the Building Regulations and be inert and non-toxic (e.g. be an approved product such as 'Dritherm').

TYPICAL SPECIFICATION FOR AN EXTERNAL WALL

Refer to clauses 6e and 12 of Appendix B regarding standard external walls.

Being cost conscious

If you check the description that I have given for an external wall in clause 12, you will note that I have given an either/or specification. The first detailed specification has been given because merely stating that walls shall achieve 0.45 W/m^{2}K would not provide tradesman builders with any standard specification to work to whilst building the extension/dwelling and might lead to disputes. The second alternative is basically just a performance specification. The reason for allowing the builder to put forward a differing specification for the external wall is that although I prefer to use a partial fill wall with a 100 mm thermal block inner skin, there might be more cost-effective ways of complying with the Building Regulations.

Nowadays there are so many ways of providing an external wall construction that will comply with the thermal requirements of the Building Regulations that I do not wish to prevent builders from using cheaper approved alternatives.

OTHER BUILDING CONTROL REQUIREMENTS – STABILITY OF
EXTERNAL CAVITY WALLS

Within Approved Document A, pages 13 to 30 set out the rules that must be
observed when designing walls. I would suggest that you obtain a copy of
Approved Document A and study it. It must be borne in mind that any set of
rules is open to interpretation. At times, one rule can conflict with another.
When designing walls/extensions/full houses, some of the most important
points to bear in mind are:

(a) On page 17 of Approved Document A it states that the information only
     covers residential buildings not more than three storeys high.
(b) The building must not exceed 15 m in height (approximately 49 ft).
(c) Floor spans – walls should not support any floor with a span larger than
     6 m (measured centre to centre of bearings.)
(d) The height of a building/extension must not be more than twice the least
     width.
(e) On page 18, Diagram 9 of Approved Document A, the maximum floor
     area of a room enclosed by structural walls is 70 m^2 or when open on one
     side 30 m^2. This is an important qualification because it places limits on
     room sizes. Once a room exceeds those sizes then the skills of an engineer
     are needed.
(f) On page 15 of Approved Document A it states that all cavity walls must
     have internal skins at least 90 mm thick and 50 mm cavities. (**Note:** the
     details indicated in this book comply with these rules as they are shown
     with 100 mm inner skins.)
(g) A thicker upper wall cannot be supported on a thinner lower wall. As an
     example, unless you have an engineer who can provide designs for, say,
     steel supports, you cannot build a cavity wall upstairs on a half brick wall
     to, say, an existing garage downstairs. The garage has to have cavity walls
     below.
(h) Using Diagram 15 of Approved Document A (see also Fig. 14.2), where
     an internal partition wall acts as a buttress to an outer cavity wall, door
     openings in the partition have to be at least 550 mm away from the outer
     wall otherwise the partition will not be treated as a buttressing wall.
(i) You are advised to study Diagram 17 of Approved Document A because
     what I am about to state does not cover all circumstances, but in general
     terms, piers between windows and doors in external walls must be at least
     a sixth of the combined width of the openings. For example, if there are
     two windows in a wall each 1.20 m wide, the pier between them must be
     at least $(1.20 + 1.20)/6 = 0.4$ m $= 400$ mm (Fig. 14.3(b)).
(j) Diagram 17 of Approved Document A indicates that a wall cannot just
     end. It has to be supported by a buttressing wall (unless calculations are
     provided to prove that the wall is stable). Where a window or door is
     sited near a corner of a building/extension, using Diagram 17, I interpret

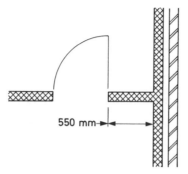

Fig. 14.2 Door opening.

the regulations as meaning that the minimum dimension between the external corner and the side of the window/door has to be a minimum of 385 mm plus the wall thickness. Based upon the wall details indicated in this book, for a wall composed of 100 mm blockwork, 50 mm cavity and nominal 112 mm brickwork, the minimum dimension would be 385 mm + 262 mm = 647 mm (Fig. 14.3(a)).

(k) Unless calculations are provided to prove otherwise, openings in walls cannot exceed 3 m in length (Fig. 14.3(b)).

THE USE OF RESTRAINT STRAPS/LATERAL SUPPORT

In older properties, bulges in external walls are very often noticeable. Although corroding cavity ties are mainly to blame, lack of lateral support is another cause. In theory, the floor and ceiling joists of a house act as ties and hold the walls in position but the number of failures has proven that the walls do move away from the joists, especially where walls run parallel to the joists or where a staircase runs alongside a wall.

The solution to the problem is the incorporation of metal restraint straps which ensure that walls, floors and roofs are secured together to prevent movement. Pages 27 and 28 of Approved Document A detail the requirements for lateral support. The basic principle is that flat roofs, first floor joists, ceiling joists and roof timbers should be strapped to the walls at centres not exceeding 2 m with galvanized mild steel (stainless steel is better but very expensive), 30 mm × 5 mm cross-section. Where joists run parallel to the wall, the restraint straps run at right angles across the joists.

WINDOW AND DOOR OPENINGS/ARCHES/LINTELS

To comply with the new 1995 thermal regulations, if you are using the elemental method (page 8 of Approved Document L) clause 1.4 indicates that

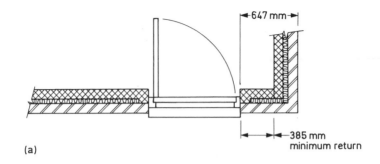

(a)

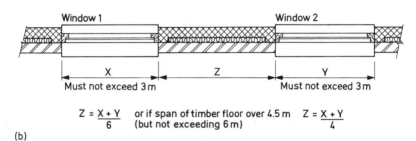

Window 1   Window 2

X
Must not exceed 3 m

Z

Y
Must not exceed 3 m

$Z = \dfrac{X + Y}{6}$ or if span of timber floor over 4.5 m $Z = \dfrac{X + Y}{4}$
(but not exceeding 6 m)

(b)

Fig. 14.3(a) External corner; (b) Pier between openings. (Details after Approved Document A.)

windows, doors and rooflights together should not exceed 22.5% of the total floor area. Clause 1.5 amplifies this by saying that in extensions the U-value of the doors, windows and rooflights should not exceed 3.3 W/m²K. Table 2 then details various values of typical windows. Under the new regulations, double glazing is taken as standard.

The floor area of the extension and existing can be used to work out the percentage of windows to floor area. Table 3 allows the percentage of windows, doors and rooflights to floor area to be varied from 22.5% if triple glazing or enhanced double glazing such as Pilkington's K glass is used.

If you refer to Fig. 14.1 you will note the standard cavity closing details around doors and windows. It is now a requirement (see page 11 of Approved Document L) that thermal bridging is avoided.

The most usual type of lintel used on modern extensions is metal 'Catnic' or similar type (Fig. 14.4). These lintels/beams are made from sheet steel that has been galvanized or treated against rusting, come in a wide variety of forms and have been designed by the manufacturer to take normal loads above door and window openings. If you are in any doubt about the strength of any lintels specified, most manufacturers will check the loads for you free of charge if you send them a copy of your plan. (Make sure that there is no charge prior to sending the plan because company policies can change from time to time.)

(a)

Typical application – lintel
over double doors in 100mm
load bearing wall

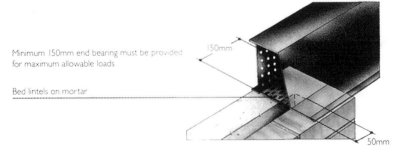

Minimum 150mm end bearing must be provided
for maximum allowable loads

150mm

Bed lintels on mortar

50mm

Lintel to extend 50mm
(minimum) beyond
reveal closer

(b)

Fig. 14.4 Steel lintels. (Reproduced with permission of Catnic Lintels Ltd.)

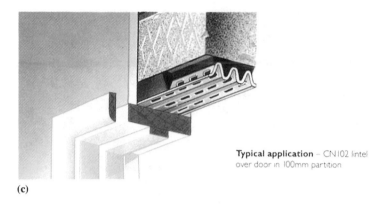

Typical application – CN102 lintel
over door in 100mm partition

(c)

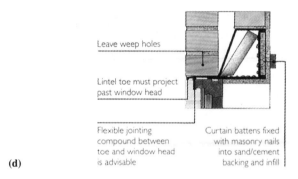

Leave weep holes

Lintel toe must project
past window head

Flexible jointing
compound between
toe and window head
is advisable

Curtain battens fixed
with masonry nails
into sand/cement
backing and infill

(d)

Fig. 14.4 contd.

Adequate bearing at the ends of lintels is essential (Fig. 14.4(b)). The manufacturer's recommendations should be followed but the usual bearing is 150 mm (6 in) each end. If you don't give enough end bearing, the lintel will fail under load. The advantage to the designer of using any of the standard lintel/beam types is that Building Control will not usually require calculations to prove the lintels (unlike RSJs, see below for details) as long as the manufacturer's load tables are followed. Although I have provided some information on these types of lintels, try to obtain manufacturers' catalogues and study the details in greater depth.

Where one is forming an opening in an existing wall to connect the new extension to the existing, it is normal practice to install two RSJs (rolled steel joists) or steel beams bolted together and seated on adequate concrete padstones. The 'Catnic' type beam can also be specified, but the reason why I normally avoid patent beams for inserting into old walls is simple practicality. Any beam being inserted into an existing wall tends to be 'bashed about a bit'. It would be difficult for even the heaviest handed building operative to

```
BEAM CALCULATION FOR  EXAMPLE CALCULATION
BEAM SPAN IS  3 METRES
BEAM IS R.S.J's
BEAM SIZE IS  178 x 102
STEEL GRADE IS 43
SLENDERNESS RATIO IS 136.00

WEIGHT CALCULATION FOR LOAD 1
ROOF LOAD          3.00  x   3.00  x  95.00   KG              =     8.39 KN
SNOW LOAD          3.00  x   3.00  x   0.75   KN  live        =     6.75 KN
CEILING LOAD       2.80  x   3.30  x  86.00   KG              =     7.80 KN
FLOOR WEIGHT       3.00  x   1.20  x  30.00   KG              =     1.06 KN
LIVE FLOOR LOAD    3.00  x   1.20  x 150.00   KG  live        =     5.30 KN
WALL LOAD          2.60  x   3.00  x 244.80   KG              =    18.73 KN
ROOF EXT           3.00  x   0.75  x  95.00   KG              =     2.10 KN
SNOW LOAD          3.00  x   0.75  x   0.75   KN  live        =     1.69 KN
                   0.00  x   0.00  x   0.00   KN              =     0.00 KN


LOADING STRESSES ARE:-
LOAD 1
TOTAL LOAD IS          52.43571 KN
LIVE LOAD IS           13.73488 KN
BENDING MOMENT IS      WEIGHT x LENGTH   =       19.66339  KNM
                              8

SHEAR STRESS IS           WEIGHT          =       27802.6  KN/M²
                       2 x WEB AREA

DEFLECTION IS         WEIGHT x LENGTH³    =       1.513737  mm
                 76.8  x ELASTIC MODULUS x MOMENT OF INERTIA

TOTAL BENDING MOMENT IS            19.66 KNM
MAXIMUM ALLOWED BENDING MOMENT =   23.26 KNM
TOTAL SHEAR STRESS IS              27802.60 KN/M²
MAXIMUM ALLOWED SHEAR STRESS IS    100,000.00 KN/M²
TOTAL DEFLECTION IS                1.51 mm
ALLOWED DEFLECTION IS              8.33 mm

Padstone dimensions
To produce a compressive stress of      0.28  N/mm²
The Padstone area should be           93635  mm²
```

Fig. 14.5 Printout from 'Pocket Engineer'. (Reproduced with permission of WL Computer Services.)

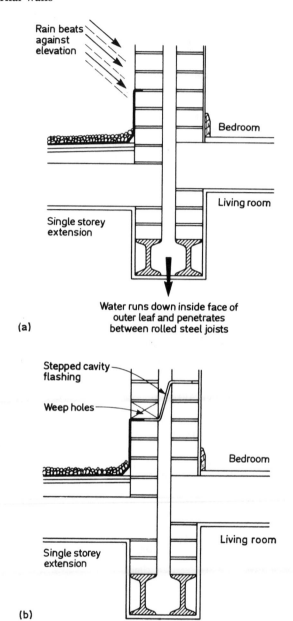

Rain beats
against
elevation

Bedroom

Living room

Single storey
extension

Water runs down inside face of
outer leaf and penetrates
between rolled steel joists

(a)

Stepped cavity
flashing

Weep holes

Bedroom

Living room

Single storey
extension

(b)

Fig. 14.6 Cavity flashing detail for steel beams: (a) incorrect; (b) correct.

seriously damage a heavy steel RSJ, but a thin fabricated lintel like a 'Catnic' might well suffer if subjected to rough treatment.

Where RSJs are being used in an old cavity wall, for most openings up to 3 m wide, in two-storey properties, it is usual to insert two independent 178

× 102 mm (still commonly known as 7 in × 4 in) RSJs (or beams of a similar size) and then bolt them together through the webs.

However, it is unwise to assume that two 7 in × 4 in RSJs (or beams of a similar size) will always carry the load. In some circumstances larger RSJs or beams may be required. If there is any doubt, you should calculate the loads and make sure that the sizes specified are adequate. Besides, more than likely the Building Control Department will require calculations anyway.

Describing structural calculations is beyond the scope of this book, but the calculations are not difficult. You therefore have several alternatives if calculations are required, which are as follows:

(a) Ask your structural engineer to design a beam for you and to provide calculations.

(b) Inform the Building Control Department that you require a conditional approval on the steelwork element. Conditional approval should be requested where a design either cannot be totally proven at plan submission stage or where the designer does not have the expertise to provide the information. (The plans will then be approved subject to the Building Control Officer receiving the missing information prior to commencing work.)

(c) Move away from the RSJ approach altogether and use a patent beam. If required, manufacturers will normally provide designs for their patent lintels if you send them a copy of your drawing. However, the provision of calculations is usually conditional upon receiving a firm order and therefore, as with (b) above, your application to Building Control would need to ask for conditional approval with regard to the beams.

(d) Obtain a suitable book from your local library and carry out the calculations yourself.

(e) Use a computer. A good program, 'Pocket Engineer', is produced by WL Computer Services (0151 426 7400). It is only a very simple program and is only designed for use with domestic situations but does provide hardcopy printouts which can be sent to Building Control as proof that the beams in question are satisfactory. The program runs on IBM and most Acorn computers (e.g. BBC Master), but some reading around the subject is advisable if you have no knowledge of the theory of how a steel beam works (Fig. 14.5 shows a typical printout).

Where RSJs are used in a cavity wall, it is essential that a proper cavity flashing is installed, otherwise water in the cavity could damage finishes in the extended property (Fig. 14.6 a and b).

Whilst dealing with RSJs, there is a large amount of useful information in Chapter 6.5 of *NHBC Standards* (Volume 2) called 'Steelwork supports to upper floors and partitions'.

15 Internal walls

GENERALLY

The internal walls of a house serve three main purposes:

(a) To subdivide the property into separate rooms.
(b) To provide support for the outside walls or the roof.
(c) In a semidetached or terraced property, to subdivide and provide fire and sound protection between dwellings.

In most new properties the construction of partition walls is either:

(a) Blockwork (sometimes brickwork if additional strength is required).
(b) Stud partitioning/patent plasterboard partitions.

BLOCKWORK PARTITIONS

It was common practice in the not too distant past to build load-bearing internal blockwork partitions off the floor slab and, other than a slight thickening in the slab, not to bother with a proper foundation. As blockwork is usually far heavier than studwork, and as other loads from the roof and upper floors were being transmitted down onto the lower floor slab, the inevitable result of this practice was that cracks developed. In bad cases of overloading, the internal walls also developed cracks as the weakened floor gave way.

In my experience, nowadays, most Building Control Departments query the use of slab thickening (unless engineers' calculations are supplied) and require the blockwork partitions (whether load-bearing or otherwise) to be taken down to a proper foundation similar to the external walls. As the blockwork in internal partition walls does not usually carry the same sort of loading as the external walls, and as the blockwork in partitions is normally only 100 mm thick, the footing can be slightly narrower than a footing that supports a cavity wall.

Despite the fact that Approved Document A might allow a narrower foundation if the ground conditions/loading were appropriate, in my opinion, (some might disagree), a practical minimum size of a strip footing for a 100 mm partition wall to a domestic extension would be 500 mm because a bricklayer needs to be able to work in the trench. Like the external walls, brick and block partition walls have a damp-proof course which is lapped into the damp-proof membrane in the ground floor slab. Where blockwork is used as partitioning between rooms of a house, there is no need to use thermal blockwork, which is normally more expensive than non-thermal blocks.

As indicated in the previous chapter, internal partition walls very often act as buttresses for the outer walls. Stud partitions (see below) should not be specified to act as buttress walls. (This does not apply to purpose-made timber frame construction that has been designed by an engineer.) Sometimes it is also necessary to provide blockwork partitions on lower floors merely to provide support for, say, a purlin in the roof space. As the purlin load has to be carried down to the ground, a supporting partition wall normally has to run through the entire height of the building.

STUD PARTITIONS

Stud partitions are basically a timber framework (either 50 mm × 75 mm or 50 mm × 100 mm) with plasterboard either side. For strength against impact, the plasterboard each side should be 12.5 mm thick. Stud partitions tend to be used extensively on upper floors of modern housing because being very light in weight, they can be built off the upper floor deck and normally do not have to follow the line of the walls below.

When building off an upper floor using stud partitions, the joists directly under a stud partition should be doubled (when the partition is built in the direction of the joists). Generally speaking, when designing small extensions, it is unwise to attempt to use stud partitions as load-bearing walls. If support is needed for the upper floors or in a roof space, a brick or block wall should be indicated. (These comments do not apply to purpose-made timber frame construction that has been designed by an engineer.)

PARTY WALLS

Party walls between two semidetached properties/terraced properties do not require the same standard of thermal insulation as an external wall. Sound transmission, or the deadening of it, is the most important factor. Approved Document E of the Building Regulations sets out schedules of wall constructions that will comply with the requirements. The basic principle is that sound transmission is deadened by using heavy materials.

As an example, if solid 215 mm blocks were used the block density would have to be 1840 kg/m^2 with 13 mm of plaster each side. The party wall must be carried up into the roof space and up to the underside of the slates/tiles.

The top of a party wall has to be fire stopped and this stopping will normally suffice for sound insulation purposes.

STEEL BEAMS

Refer to previous chapter.

16 Ground floors

Ground floors to domestic extensions can be divided into three basic types:

(a) Solid
(b) Patent concrete (not detailed)
(c) Timber.

Figures 16.1 to 16.4 show the general construction of these types of floors.

COMMON FAULTS WHEN DEPICTING THE GROUND FLOOR CONSTRUCTION ON A DRAWING

When preparing a plan for an extension/dwelling several important details must be observed:

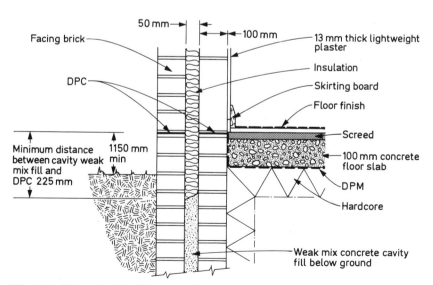

Fig. 16.1 Uninsulated solid ground floor.

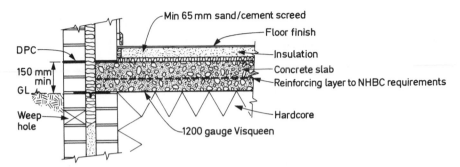

Fig. 16.2 Typical NHBC suspended floor slab (used where hardcore under floor exceeds 600 mm thickness).

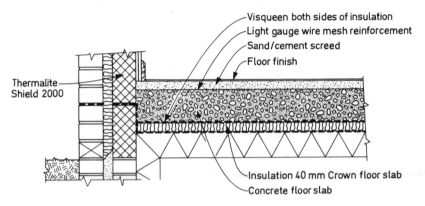

Fig. 16.3 Insulated solid ground floor.

(a) Do not show the surface level of finished ground floor slab or subfloor surfaces as being lower than external ground level. The house/subfloor could flood in wet weather.

(b) Where the hardcore beneath a solid floor exceeds 600 mm in thickness you are advised to refer to *NHBC Standards* (Volume 1). NHBC terms any concrete ground floor with more than 600 mm fill as being a suspended floor and details how the slab thickness and reinforcement should be calculated (Figure 16.2 shows a typical NHBC suspended floor slab – note how the slab is built on to the internal skin of the cavity wall for additional support.) Alternatively where there are large subfloor voids, consider either the 'Housefloor' or suspended timber floor options (the latter is detailed later in this chapter).

(c) Approved Document L (1995 Edition) has introduced a wide number of changes to the insulation standards required in most types of floor and this is discussed later in this chapter (on page 123).

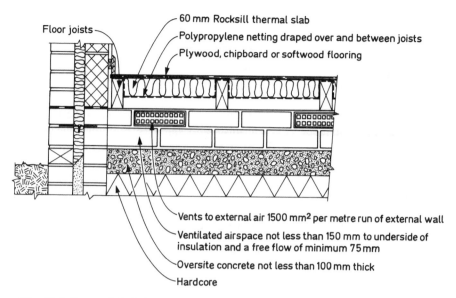

Floor joists

60 mm Rocksill thermal slab

Polypropylene netting draped over and between joists

Plywood, chipboard or softwood flooring

Vents to external air 1500 mm² per metre run of external wall

Ventilated airspace not less than 150 mm to underside of insulation and a free flow of minimum 75 mm

Oversite concrete not less than 100 mm thick

Hardcore

Fig. 16.4 Suspended timber ground floor.

SOLID GROUND FLOORS IN CONTACT WITH THE GROUND

The most common type of floor construction being used in modern extensions/dwellings in England and Wales is the solid floor. Fig 16.1 shows an uninsulated floor. In buildings exempt from Building Control (e.g. conservatories, porches and the like), this type of floor can still be used. It should also be used in attached garages where the floor will be subjected to heavy loads. There are other places where this type of construction may also be used in habitable areas. As I have mentioned before, extensions under 10 m² to older properties are not subject to the same controls as larger extensions (see clause 0.3 of Approved Document L).

So, in theory, the uninsulated floor can still be used on very small extensions built prior to the introduction of any form of thermal regulation. In my experience, Building Control Officers usually ask for an insulated floor in all new habitable areas. If you feel like arguing the point, you can sometimes convince the officer concerned that insulation is not required. In practice, it is far easier to agree to insulating the floors of all habitable areas.

Figures 16.2 and 16.3 show the modern solid floors complete with a thermal insulation layer in position. Figure 16.2 shows an NHBC suspended floor. This type of construction should be used if the hardcore fill under the floor exceeds 600 mm in thickness.

Refer to clause 9 of Appendix B for standard floor specification. Note in particular the insulation layer that is incorporated. From enquiries made with the manufacturers, Owens Corning, the layer specified should make the floor comply with the new thermal insulation standards.

PATENT SUSPENDED FLOORS

In recent years, the 'Housefloor' system has been introduced to the UK and comprises purpose-made concrete joists and standard building blocks (not detailed here because it is unlikely to be used in small domestic extension).

SUSPENDED TIMBER FLOOR

The timber ground floor (see **Fig. 16.4**) still requires:

(a) A hardcore bed and 100 mm concrete slab as with the solid floor, or
(b) 50 mm of concrete or inert fine aggregate and a polythene damp-proof membrane.

In large floor areas, honeycomb sleeper walls usually support the ground floor joists which reduces the spans and cross-sections of joists being used in the floor. (Honeycomb sleeper walls are walls built with air holes in them to provide good through ventilation beneath the floor.)

Obviously, where honeycomb sleeper walls are required, the alternative (a) above is used to provide a firm foundation for these walls. As alternative (b) only has limited application, it is obvious that the construction below a timber floor is very similar to a solid floor. As there is a duplication of materials below a timber ground floor, it makes this form of construction more expensive than a solid ground floor under normal conditions. (**Note:** If a large amount of fill is required below a solid floor because of steeply sloping site levels, it could be cheaper to put in a timber floor.)

Between the honeycomb walls and the joist, it is normal to incorporate a wall plate which is bedded on a DPC. Once again, when drawing your plan, keep the concrete subfloor surface above ground level. The voids underneath timber floors have to be provided with ventilation to the outside air. This is done by installing air bricks, but they should not be built in too low, otherwise water will flood into the house in wet weather. The purpose of air bricks beneath a timber floor is to prevent dry rot spores germinating on the wood of the floors. The ventilation should be equal to 1500 mm^2 per metre run of external wall.

Note on dry rot: Dry rot is a fungus that literally eats wood and it has a terrific growth rate. To germinate, true dry rot needs three things:

(a) Moisture in the wood over 20%.
(b) Bad ventilation.
(c) Warmth.

The provision of air bricks prevent the conditions arising where dry rot can flourish. If you specify tanalized timber this will also help to stop wood-rot outbreaks.

Refer to clause 11 of Appendix B for standard floor specification. Note in particular the insulation layer that is incorporated. From enquiries made with the manufacturers, the thickness specified should make the floor comply with the new thermal insulation standards.

17 Timber upper floors

Figures 17.1 and 17.2 show the general construction of timber upper floors. Except where built over partially ventilated spaces such as garages, timber upper floors do not require insulation. Unlike suspended timber ground floors, the joists of upper floors have to be of a much larger section because the spans cannot be broken by means of such things as sleeper walls. Joist sizes can be obtained from Approved Document A. The normal centres for joists used on upper floors is 450 mm. The normal procedure that I adopt when designing a timber upper floor is to find the longest span and make all the joists the same size. The reason for this overdesign is cosmetic. If you do not keep all the sections the same, the floor to ceiling heights in each room will vary considerably and this can look ridiculous. Where stud partitions are

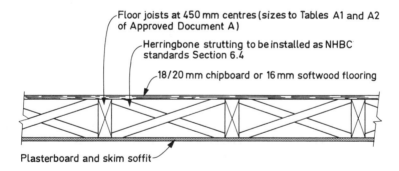

Floor joists at 450 mm centres (sizes to Tables A1 and A2 of Approved Document A)

Herringbone strutting to be installed as NHBC standards Section 6.4

18/20 mm chipboard or 16 mm softwood flooring

Plasterboard and skim soffit

Fig. 17.1 Cross-section of timber-joisted first floor.

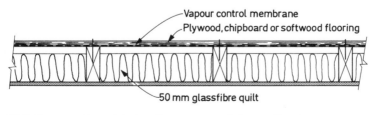

Vapour control membrane

Plywood, chipboard or softwood flooring

50 mm glassfibre quilt

Fig. 17.2 Cross-section of timber floor over partially ventilated space (e.g. garage).

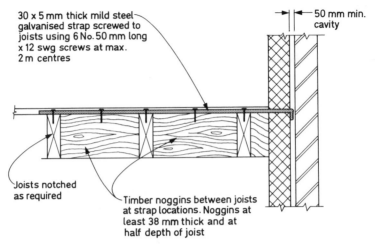

30 x 5 mm thick mild steel galvanised strap screwed to joists using 6 No. 50 mm long x 12 swg screws at max. 2 m centres

50 mm min. cavity

Joists notched as required

Timber noggins between joists at strap locations. Noggins at least 38 mm thick and at half depth of joist

Fig. 17.3 Restraint straps with joists parallel to wall.

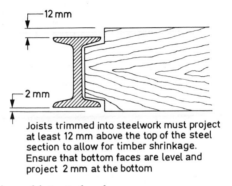

12 mm

2 mm

Joists trimmed into steelwork must project at least 12 mm above the top of the steel section to allow for timber shrinkage. Ensure that bottom faces are level and project 2 mm at the bottom

Fig. 17.4 Joists trimmed into steelwork.

built off upper floors, the floor beneath each partition running with the joists by doubling up the joists. Doubling up also is necessary under baths. Upper floors can be covered with tongued and grooved boarding or more commonly nowadays, flooring grade chipboard. The NHBC recommend 16 mm minimum thick floor boards or 18/20 mm minimum chipboard type C4.

If you refer to Fig. 17.3, you will note that the floor joists should have restraint straps fixed to them to provide lateral support to external walls. Figure 17.4 indicates how a steel beam should be fitted when supporting a timber floor.

18 Flat roofs

TYPES OF FLAT ROOF

There are basically three designs of flat roof:

(a) Cold deck (Figs. 18.1 to 18.2)
(b) Warm deck (Fig. 18.3)
(c) Inverted warm deck (not detailed).

COLD DECK

The cold deck has the insulation laid between the joists and the space above is ventilated to the outside air in order to prevent moisture condensing in the cold roof void.

WARM DECK

With the warm deck, the insulation is placed on top of the roof deck and the void between the deck and the plasterboard ceiling stays warm because it is protected by the insulation. On a warm deck roof, ventilation should not be provided to the enclosed voids.

FLAT ROOF STRUCTURE GENERALLY

If you refer to Figs. 18.1 to 18.3 you will note that whether you are dealing with cold deck or warm deck construction there are common elements. Starting underneath and moving upwards, these are as follows.

Plasterboard/plaster soffit

Like most modern ceilings, a flat roof normally is lined with plasterboard. As the joists will probably be at 450 mm centres, it is only necessary to use 9.5 mm thick plasterboard with a skim finish of plaster. On the cold deck system there should be a vapour check between the plasterboards and joists. This vapour check can either be created by using aluminium-foil-backed plasterboard, glassfibre with a polythene sheet bonded on or tacking a layer

NOTE :- All openings are to be continuous or provided by patent product which gives the same area of ventilation. Details after Willan Building Products (Glidevale)

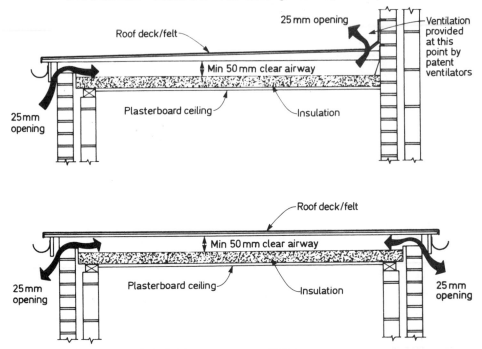

Fig. 18.1 Flat roof (cold deck): cross-sections indicating ventilation.

of 250 gauge (minimum) polythene sheeting with 150 mm laps to the joists before fixing the plasterboard. In my experience, most builders in my area who are installing a cold deck roof opt for the foil-backed plasterboard as the extra cost of the foil backing is negligible when compared with the labour cost of fixing the polythene. In a warm deck, foil-backed plasterboard must not be used as a vapour check (Figs. 18.3(a) and (b) shows vapour check position in warm deck).

Roof joists

The size of roof joists required for a flat roof can be obtained from Approved Document A or the *NHBC Standards*. These schedules indicate timbers of either grade SC3 or SC4. Being a better quality of timber, SC4s will span slightly further that SC3s but are obviously more expensive. When dealing with warm deck roofs, the joist size can be taken from the tables. However, with cold deck roofs, it is a requirement of the Approved Documents that there is a free ventilation space of 50 mm minimum left above the insulation layer. If the lowest *U*-value allowed for a flat roof (0.35 W/m^{2}K) is adopted for the roof design the normal insulation used is 150 mm of glassfibre. As the insulation goes between the joists the minimum height of joists has to be 150 mm (plus 50 mm thick cross-battening to provide a 50 mm airspace above the quilt) or 200 mm if adequate cross-ventilation can be provided by using joists on their own.

Firrings

Falls (slopes) are created by installing firrings on top of the roof joists (which should be laid horizontally) and below the plywood or chipboard roof deck.

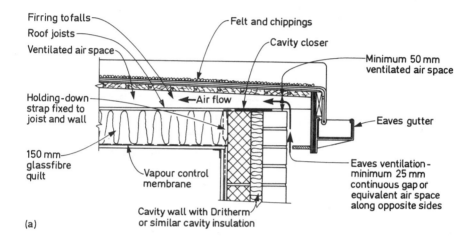

Firring to falls
Roof joists
Ventilated air space

Felt and chippings

Cavity closer

Minimum 50 mm ventilated air space

Holding-down strap fixed to joist and wall

←Air flow

Eaves gutter

150 mm glassfibre quilt

Vapour control membrane

Eaves ventilation – minimum 25 mm continuous gap or equivalent air space along opposite sides

Cavity wall with Dritherm or similar cavity insulation

(a)

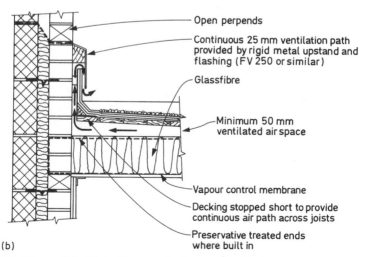

Open perpends

Continuous 25 mm ventilation path provided by rigid metal upstand and flashing (FV 250 or similar)

Glassfibre

Minimum 50 mm ventilated air space

Vapour control membrane

Decking stopped short to provide continuous air path across joists

Preservative treated ends where built in

(b)

Fig. 18.2 Flat roof (cold deck) – ventilation details: (a) eaves; (b) joists built into external wall. (Details after Owens Corning Insulation Ltd.)

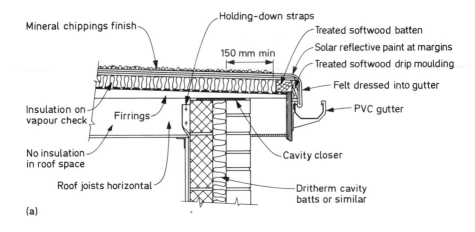

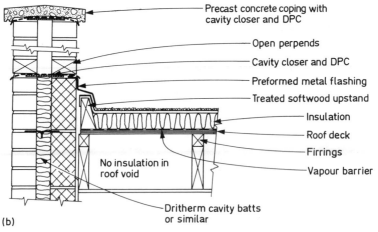

Fig. 18.3 Flat roof (warm deck) – vapour check details: (a) eaves; (b) joists built into parapet wall. (Details after Owens Corning Insulation Ltd.)

Firrings are shaped timber battens that start wide at one end and then taper down uniformly to create a regular fall and/or cross-fall on the roof deck. The NHBC recommend the following minimum (measured at the thinnest end) sizes of firring:

- Joists at 450 mm centres or below – 38 mm × 38 mm.
- Joists above 450 mm up to 600 mm centres – 38 mm × 50 mm deep.

The roof deck

Supporting the finish there is the roof deck (usually roofing grade chipboard (type C4) or pretreated WBP plywood). If the client is agreeable to additional cost, a plywood base to the roof is far superior to a chipboard one.

The NHBC states that if chipboard is used for the roof decks it must be type C4. For ease of recognition on site, type C4 has colour stripes (red and green) applied to the edges at time of manufacture. It is important that builders do not use an inferior grade as the lesser grades do not have adequate moisture resistance when used in roof decks and will inevitably crumble.

Roof straps

In recent hurricane force winds that swept the UK, some flat roofs literally took off and ended up in the back garden. In order to combat wind lift, it is now a requirement of the Building Regulations that all roofs are strapped down to the surrounding walls using 30 mm × 5 mm galvanized mild steel straps at 2 m centres. (**Note:** The NHBC require the straps to be at 1.20 m centres.)

Roof finishes

The most commonly used roof finishes on a flat roof are:

(a) Bitumen felt, usually covered by a 13 mm layer of white limestone chippings laid shoulder to shoulder in bitumen. The purpose of the chippings is to reflect the heat of the sun. Without them the felts could very quickly perish.
(b) Asphalt.

Gutterwork and downpipes

Approved Document H of the Building Regulations sets out the rules for the size of gutters and downpipes. Assuming only one downpipe and one length of guttering the following sizes are needed:

(a) For roofs of 18 m^2 the gutter has to be 75 mm (min.) with an outlet of 50 mm (min.).
(b) For roofs of between 18 and 37 m^2 the gutter has to be 100 mm (min.) with 63 mm (min.) outlet.

All rainwater pipes should discharge into gullies and from there to underground drainage. In order to ensure that water does not splash up from the gullies most builders connect the downpipes into back inlet gullies (BIGs).

FLAT ROOFS VERSUS PITCHED ROOFS

There is little doubt in terms of life expectancy, that a tiled/slate roof (pitched roof) is far superior to a flat roof. The most common faults with a flat roof are as follows:

(a) Making flat roofs totally without any drainage slopes was probably the main reason for there being so many leaks associated with flat roof construction in the past. The term 'flat roof' is inaccurate because a flat roof made from standard BS 747 felt should **never** be flat. They should have a slight fall on them (minimum 1:40).

(b) If 'falls' were provided they were not large enough and water 'ponded' on the surface and eventually worked its way through the membrane. In order to drain satisfactorily the deck on domestic extensions should be laid to falls of minimum 1:40. One also has to remember that on large flat roofs it will be necessary to create 'falls' and 'crossfalls' (i.e. the roof water needs to be directed to a drainage point) (Fig. 18.4).

(c) The decking material used (chipboard or plywood) was the wrong grade of material for roofing and deteriorated. Chipboard should be grade C4. Plywood should be WBP grade. Chipboard should be supported at all edges by inserting additional noggins between the joists.

(d) With the cold roof, it is essential that the voids above the insulation be vented to the external air. Eaves ventilation should be provided by a gap of 25 mm left between the fascia board and the top of the wall (or an approved alternative means of ventilation installed). Condensation will form on the roof deck and will cause staining on the ceiling below and eventually rot the roof joists and roof deck if adequate ventilation is not provided.

(e) The felt finish (if applicable) has not been laid in accordance with recommended standards. When prefelted chipboard (blacktop) is used as the decking the felt layer on this chipboard must not be treated as a substitute for one of the layers of roofing felt.

(f) The surface of the felt has not been adequately protected from sunlight and as a result has flexed, bubbled and blistered.

So why use flat roofs? The simple answer is, that compared with a pitched

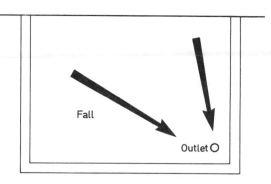

Fig. 18.4 Falls on flat roof directed to drainage point.

roof, they are comparatively cheap to construct. Maintenance is a different matter. Although flat roof construction is improving, in my experience, most flat roofs start to leak very quickly and some sort of maintenance must be anticipated within 7 to 10 years. As previously mentioned, in the recent past, we have carried out a series of designs for flat roof replacements in our area. All the flat roofs in question had only been installed a few years previously and they had all been patched many times before the owners finally decided that the only solution was to install a replacement tile/slate roof. In all cases, the flat roofs in question were leaking, and the timbers/roof decks were rotten. By comparison, a pitched roof usually has a trouble free life of at least 40 years.

After saying all that, nowadays, of the small works going through my office, barely 2% have flat roof construction. Five years ago, nearly every small extension had a flat roof.

Why the change? The answer is that Joe Public has learned the lesson the hard way. Now I do not have to try to persuade people to think long term. They want to spend the additional money and have better quality. However, there still are times when a flat roof is the only option available.

My credo is, it is best to avoid using flat roofs wherever possible.

Where it is now necessary to use flat roofs, drawings issued by my office tend to specify the cold deck type and I shall explain why.

COLD DECK VERSUS WARM DECK ON SMALL DOMESTIC EXTENSIONS

Some years back, several Building Control Departments in my area developed a new fad and tried to 'convert' our office to warm deck flat roofs for small domestic extensions. We decided to be cooperative, especially when we discovered that the NHBC did not recommend cold deck construction either.

So why do we specify cold deck? As explained before, on smaller projects of only a few thousand pounds in value, it is unusual for a surveyor to supervise the work because of the fee charges involved in such an exercise. Following our discussions with Building Control, we specified probably 10 to 15 warm deck roofs before the queries started to roll in. It was then that we began to realize that the warm deck roofs that we had specified were far from popular with tradesmen builders in our area. At first we thought it was merely resistance to change but we then made a few enquiries of our own and found that:

(a) Very few local builders' merchants were prepared to stock the warm deck roofing materials.

(b) The central distributors were not really interested in supplying small quantities.

(c) Because the quantities of warm deck insulation involved were comparatively small, it meant that the materials in question either had to be collected or that a very high delivery charge was levied by the central supplier.

(d) Where they did supply warm deck insulation, the prices being charged were high when compared to standard fibreglass rolls used in a cold deck construction.

(e) Tradesmen builders were deviating from our plans by asking, and obtaining, permission from Building Inspectors to revert to cold deck construction anyway.

We finally came to the conclusion that as flat roofs were becoming less common anyway, and as suppliers showed no signs of reducing their tarrifs, we were fighting a losing battle. Needless to say, our experiment with warm deck construction on smaller domestic extensions soon petered out. After saying all that, other surveyors in other areas may have had completely different experiences.

TYPICAL SPECIFICATION FOR FLAT ROOFS

Refer to Appendix B, clauses 6b, 6c and 13 for standard roof specification.

19   Pitched roofs

GENERALLY

The term pitched roof basically covers the standard tiled or slated roof that is the norm in the UK.

COVERINGS – TILES/SLATES

The coverings to a pitched roof comprise (Figs 19.1 to 19.5):

(a) A covering of manufactured tiles, shingles, or slates (natural or manufactured).
(b) Ridge and hip tiles to cover the exposed edges.
(c) Leadwork for flashings, soakers, aprons and the like.
(d) Battens to support the roof finish.
(e) Sarking or underfelt as a secondary barrier if a tile/slate is lost in a high wind.

The tiles or slates have to be laid to a suitable pitch (i.e. not less than the manufacturer's recommended instructions). If you refer to Figs 19.1 to 19.2 you will note that the minimum pitch is clearly indicated for each type of covering. The companies that produce roof coverings carefully test their products for weather resistance, and if their products are laid to slopes lower than the recommended minimum then there is no guarantee that they will provide the performance required. **Note**: the recommended slope must be maintained over the whole roof pitch. It is normal practice to provide tilt fillets at the eaves in order that the tiles/slates discharge into the gutters properly. I have seen cases in my locality where the builders have forgotten this principle and have inserted an overlarge tilt fillet. This created a weakness at the change in slope and the houses affected by the design fault have suffered from water penetration at the eaves.

In general terms, the smaller the tile or slate, the steeper the pitch required. In order to avoid queries by Building Control, when specifying a roof finish, you must state the pitch on your plans.

As examples, I have extracted a list of a few tiles and lowest pitch as follows:

Tile type	Min. pitch (degrees)
Marley Wessex	15
Marley Modern, Smooth	17.5
Redland Stonewold, Delta and Regent	17.5
Marley Modern, Granular	22.5
Redland Renown	30
Marley Plain	35
Redland Rosemary	40
Imitation slate type	
Marley Monarch	22.5
Bradstone Moordale	25
Bradstone Cotswold	30

As there are many manufacturers and a variety of different roof tiles/slates on the market, it would be advisable for you to contact the various manufacturers and request copies of their current catalogues. These will give you a complete performance specification and descriptions to use on your plans for each type of tile/slate. This information includes such matters as minimum head and side laps. For examples see Fig. 19.1.

Natural slates

Natural slates, whether new or secondhand, come in a variety of sizes (Fig. 19.2), and roofs will leak if the slates are too small in relation to their pitch, or slates are not wide enough in relation to their length.

Tile/slate battens

The roof tiles or slates are fixed to battens over sarking felt. With regard to sizes of battens required to support the covering, the tile or slate manufacturer's recommendations should be followed.

ROOF STRUCTURE

There are two basic types of pitched roofs now in common usage:

(a) Traditional type (Figs. 19.6 and 19.7).
(b) Trussed rafter type (Fig. 19.8).

TILES		MINIMUM PITCH*	
		GRANULAR	SMOOTH
PLAIN + MARLDEN	267 x 165 mm	65mm lap 35° — 35°	65mm lap 35° — 35°
PLAIN + MARLDEN FEATURE	267 x 165 mm	35mm lap 90° — 90°	35mm lap 90° — 90°
THAXDEN	270 x 165 mm	70mm lap 35° — 35°	
WESTWOLD	267 x 165 mm	65mm lap 35° — 35°	
LUDLOW PLUS	387 x 229 mm	75mm lap 30° — 30°	75mm lap 25° — 25° 100mm lap 22.5° — 22½°
ANGLIA PLUS	387 x 230 mm	75mm lap 30° — 30°	75mm lap 30° — 30° 100mm lap 25° — 25°
DOUBLE ROMAN	420 x 330 mm	75mm lap 30° — 30°	75mm lap 25° — 25° 100mm lap 22.5° — 22½°
LUDLOW MAJOR	420 x 330 mm	75mm lap 30° — 30°	75mm lap 25° — 25° 100mm lap 22.5° — 22½°
MENDIP	420 x 330 mm	75mm lap 30° — 30° 100mm lap 25° — 25°	75mm lap 22.5° — 22½°
MODERN + MOCK BOND	420 x 330 mm		75mm lap 22.5° — 22½° 100mm lap 17.5° — 17½°
WESSEX	413 x 330 mm		75mm lap 15° — 15°
BOLD ROLL	420 x 330 mm	75mm lap 30° — 30°	75mm lap 17.5° — 17½°
MONARCH	325 x 330 mm		75mm lap 22.5° — 22½°
EURO 9	448 x 308 mm		75mm lap 30° — 30°
PANTILE	412 x 260 mm		70mm lap 22.5° — 22½°

Fig. 19.1 Marley tile products, indicating minimum roof pitches. (Reproduced with permission of Marley Building Materials Ltd.)

Slate size (nominal) mm	inches	Moderate exposure: driving rain index less than 7 m²/s — Minimum rafter pitch									Severe exposure: driving rain index 7 m²/s or more — Minimum rafter pitch								
		20°	22½°	25°	27½°	30°	35°	40°	45°	85°	20°	22½°	25°	27½°	30°	35°	40°	45°	85°
660 × 355	26 × 14	130*	105	90	80	75	75	65	65	—	—	140	120	115	105	85	75	65	—
610 × 355	24 × 14	115	105	90	80	75	75	65	65	—	145*	125	105	95	90	75	75	65	—
610 × 305	24 × 12	130*	115	90	80	75	75	65	65	—	—	120	120	115	110	90	80	70	—
560 × 305	22 × 12	115	105	90	80	75	75	65	65	—	140*	120	105	100	90	75	75	65	—
560 × 280	22 × 11	120*	110	90	80	75	75	65	65	—	145*	130	110	105	100	85	75	65	—
510 × 305	20 × 12	115	105	90	80	75	75	65	65	—	—	125	100	95	85	75	75	65	—
510 × 255	20 × 10	125*	110	90	80	75	75	65	65	50	—	135	115	110	100	90	75	65	65
460 × 305	18 × 12	115*	105	90	80	75	75	65	65	50	—	—	110	95	85	75	75	65	65
460 × 255	18 × 10	125*	110	90	80	75	75	65	65	50	—	—	115*	110	100	85	75	65	65
460 × 230	18 × 9	125*	115*	100	80	75	75	65	65	50	—	—	120*	115*	105	95	85	65	65
405 × 305	16 × 12	—	—	—	80	75	75	65	65	50	—	—	—	—	90	80	75	65	65
405 × 255	16 × 10	—	—	—	85	75	75	65	65	50	—	—	—	—	100	95	75	65	65
405 × 230	16 × 9	—	—	—	85	75	75	65	65	50	—	—	—	—	100	100	85	65	65
405 × 205	16 × 8	—	—	—	90	75	75	65	65	50	—	—	—	—	105	100	90	65	65
355 × 305	14 × 12	—	—	—	80	75	75	65	65	50	—	—	—	—	75	75	75	65	65
355 × 255	14 × 10	—	—	—	80	75	75	65	65	50	—	—	—	—	80	75	75	65	65
355 × 230	14 × 9	—	—	—	80	75	75	65	65	50	—	—	—	—	85	80	75	65	65
335 × 205	14 × 8	—	—	—	80	75	75	65	65	50	—	—	—	—	90	85	75	65	65
355 × 180	14 × 7	—	—	—	80	75	75	65	65	50	—	—	—	—	95	90	80	65	65
305 × 255	12 × 10	—	—	—	80	75	75	65	65	50	—	—	—	—	75	75	75	65	65
305 × 205	12 × 8	—	—	—	80	75	75	65	65	50	—	—	—	—	85	80	75	65	65
305 × 150	12 × 6	—	—	—	80	75	75	65	65	50	—	—	—	—	85	80	75	65	65
255 × 255	10 × 10	—	—	—	80	75	75	65	65	50	—	—	—	—	75	75	75	65	65
255 × 205	10 × 8	—	—	—	80	75	75	65	65	50	—	—	—	—	75	75	75	65	65
255 × 150	10 × 6	—	—	—	80	75	75	65	65	50	—	—	—	—	75	75	75	65	65

Note: the actual pitch at which the slate lies on the roof is less than the rafter pitch by an amount which is a function of the slate thickness and the lap. Therefore head laps marked* are not suitable for use with extra heavy slates.

Fig. 19.2 Slate roofing dimension details. (Reproduced with permission of Alfred McAlpine Slate Products Ltd.)

D e s i g n   D e t a i l s

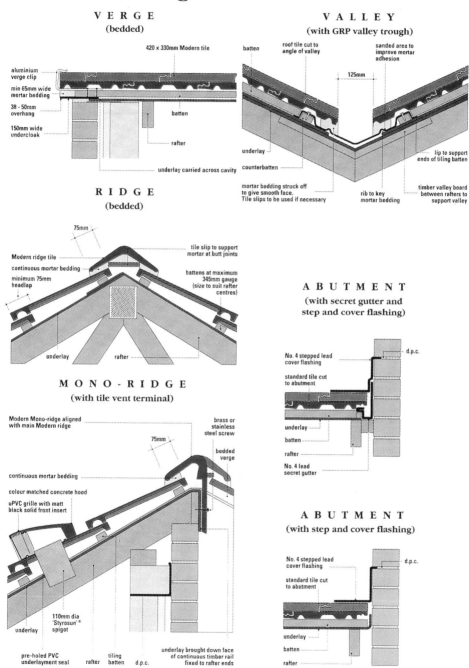

VERGE
(bedded)

420 x 330mm Modern tile

aluminium verge clip

min 65mm wide mortar bedding

38 - 50mm overhang

150mm wide undercloak

batten

rafter

underlay carried across cavity

VALLEY
(with GRP valley trough)

batten

roof tile cut to angle of valley

sanded area to improve mortar adhesion

125mm

underlay

counterbatten

mortar bedding struck off to give smooth face. Tile slips to be used if necessary

lip to support ends of tiling batten

rib to key mortar bedding

timber valley board between rafters to support valley

RIDGE
(bedded)

75mm

Modern ridge tile

continuous mortar bedding

minimum 75mm headlap

tile slip to support mortar at butt joints

battens at maximum 345mm gauge (size to suit rafter centres)

underlay

rafter

ABUTMENT
(with secret gutter and step and cover flashing)

d.p.c.

No. 4 stepped lead cover flashing

standard tile cut to abutment

underlay

batten

rafter

No. 4 lead secret gutter

MONO-RIDGE
(with tile vent terminal)

Modern Mono-ridge aligned with main Modern ridge

brass or stainless steel screw

75mm

bedded verge

continuous mortar bedding

colour matched concrete hood

uPVC grille with matt black solid front insert

110mm dia 'Styrosun'*® spigot

underlay

pre-holed PVC underlayment seal

tiling batten

rafter

d.p.c.

underlay brought down face of continuous timber rail fixed to rafter ends

ABUTMENT
(with step and cover flashing)

No. 4 stepped lead cover flashing

d.p.c.

standard tile cut to abutment

underlay

batten

rafter

Fig. 19.3 Tiling design details. (Reproduced with permission of Marley Building Materials Ltd.)

E A V E S
(with uPVC fascia and soffit)

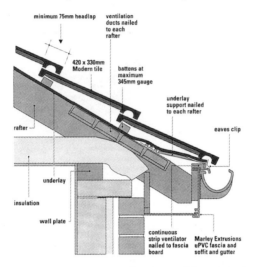

Fig. 19.3 contd.

Terminology

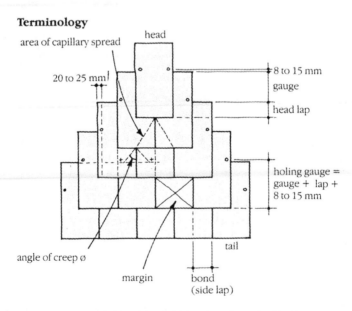

Fig. 19.4 Typical slate roof. (Details after Penrhyn Quarries Ltd.)

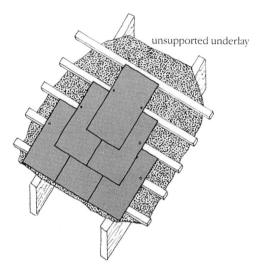
unsupported underlay

Fig. 19.5 Slate roof details.

Traditional roof

Below the roof finish (i.e. the tiles/slates and underfelt) the traditional roof comprises a framework of timbers. Each individual member has its own traditional name (Figs 19.6 and 19.7). Sizes of roof members required within a roof can be obtained from Approved Document A of the Building Regulations or from *NHBC Standards*.

Their common names and their functions are:

(a) **Rafters** (common rafters) – these are usually spaced at 450 mm centres and carry the weight of the roof finish.
(b) **Ceiling joists** – for roofs of any reasonable span, the ceiling joists perform two functions –
    (i) to hold up the ceiling;
    (ii) to act as ties and prevent the roof spreading by forming the third side of the triangle.
(c) **Purlins** – heavy timber beams that support the rafters. They are not always needed on smaller roofs.
(d) **Hangers, binders** and **struts** – subsidiary members that help to further triangulate the roof and strengthen it.
(e) **Ridge board** – this member gives stability to the rafters and also a fixing for the cut ends of common rafters.
(f) **Hip rafter** – a large rafter that runs underneath a hip on a hipped roof and carries the cut ends of common rafters.
(g) **Valley rafter** – a large rafter that runs underneath a valley on a roof and carries the cut ends of common rafters.

(h) **Wall plate** – a horizontal timber bedded onto the top of the brickwork to ensure that the top of the wall is level.

Note: One of the most glaring defects of the last few editions of the Building Regulations is that they all ignore elements (f) and (g) above. The NHBC also appear to have dodged the issue and have placed roof design, with regard to hips and valleys, into the hands of an engineer.

Appendix 7.2 of *NHBC Standards* (Volume 2) is of some help, even though it does not provide the complete answer. The NHBC's minimum requirements are as follows:

- Struts and braces – 100 mm × 50 mm.
- Hips – Minimum size should be the width of the cut end of the rafter plus 25 mm (no width is given).
- Valleys – minimum size should be 32 mm thick. (No depth is given.)
- Ridge board – no thickness given. Minimum size should be width rafter cut plus 25 mm.

In the past, when queries have been raised by various Building Control Departments concerning sizes of these various elements, and I have tried to substantiate the sizes that I have given, the only simple guidelines that anyone seems to be able to unearth is information contained within a very

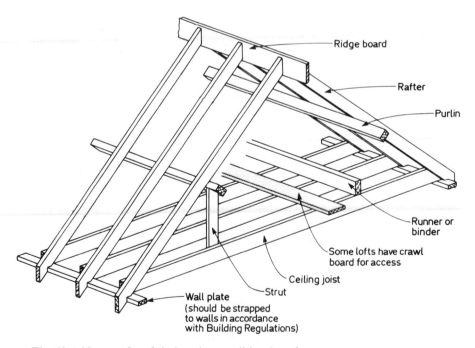

Fig. 19.6 Names of roof timbers in a traditional roof.

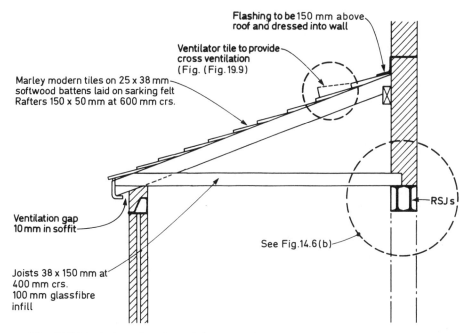

Flashing to be 150 mm above roof and dressed into wall

Ventilator tile to provide cross ventilation (Fig. (Fig. 19.9)

Marley modern tiles on 25 x 38 mm softwood battens laid on sarking felt
Rafters 150 x 50 mm at 600 mm crs.

RSJs

Ventilation gap 10 mm in soffit

See Fig.14.6(b)

Joists 38 x 150 mm at 400 mm crs.
100 mm glassfibre infill

Fig. 19.7 Typical section through lean-to extension showing pitched roof over 15°.

outdated copy of some old building byelaws (believed to be circa 1936) which gives sizes on roofs of 35° and 45° pitches which are as follows:

Valley rafters

(a) Make 1.50 in (40 mm) thick and twice the depth of common or jack rafter where the valley is supported intermediately by purlins which latter are strutted under junction with valley rafter, or
(b) make 2 in (50 mm) thick and same depth as specified for ridges and hip rafters where valley rafter is intermediately supported as (a).

Note: valley rafters are beams subject to considerable loads and adequate strength is essential.

Hip rafters and ridges (dimensions in inches (mm))

Depth of rafters	Roof pitch up to about	
	35°	45°
4 (100)	7 × 1.25 (175 × 32)	8 × 1.25 (200 × 32)
5 (125)	8 × 1.25 (200 × 32)	9 × 1.25 (225 × 32)
6 (150)	9 × 1.25 (225 × 32)	11 × 1.25 (300 × 32)

Note: The figures in brackets are my approximate metric conversions. It should be borne in mind that some Building Control Officers may not be prepared to accept the use of old tables.

I normally indicate that wall plates should be 100 mm × 50 mm.

Trussed rafter roof

With trussed roofs, the rafters and other elements used in a traditional roof are substituted by factory designed and manufactured roof trusses. In order that a trussed rafter roof can be passed by Building Control, you must obtain calculations from the truss manufacturers to support your application. As most truss manufacturers will not provide calculations until they receive a firm order, you can ask Building Control to approve the plans on the understanding that the calculations will be provided before the roof work starts on site (a conditional approval).

Trussed rafters cannot be made on site (unless the contractor has the necessary plant – which is very unlikely.) Trusses are factory made components, manufactured to BS 5268. Trussed rafters are supplied in a variety of types (Fig. 19.8 gives the names of the most common trussed rafter types). The structure of a trussed rafter roof comprises the following elements:

(a) The roof trusses which are usually spaced at 600 mm centres.

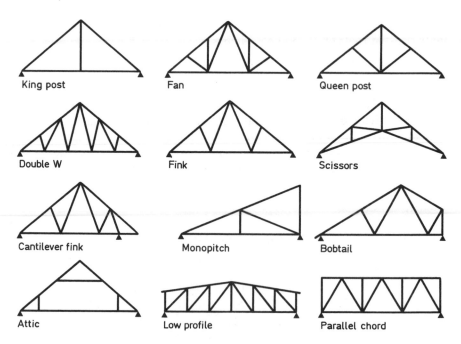

Fig. 19.8 Standard trussed rafter configurations.

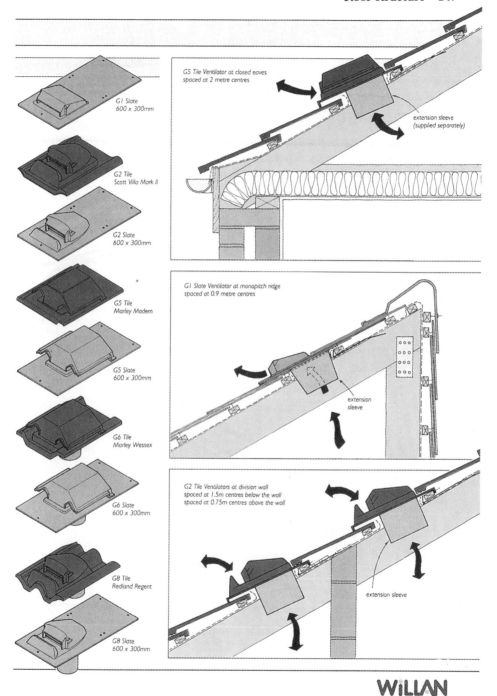

Fig. 19.9 Types of tile and slate ventilators. (Reproduced with permission of Willan Building Services (Glidevale Ltd).)

(b) Timber bracing (minimum 100 mm × 25 mm cross-section) which stabilizes the roof trusses. This bracing must not be installed in a haphazard fashion. Appendix 7.2-E of *NHBC Standards* (Volume 2) indicates the requirements for various roof layouts.

INSULATION AND VENTILATION

See Appendix B, clause 6a for standard specification. By my calculations it will now be necessary to incorporate 160 mm of glassfibre or other approved insulant between and over the ceiling joists. In addition, the void above the

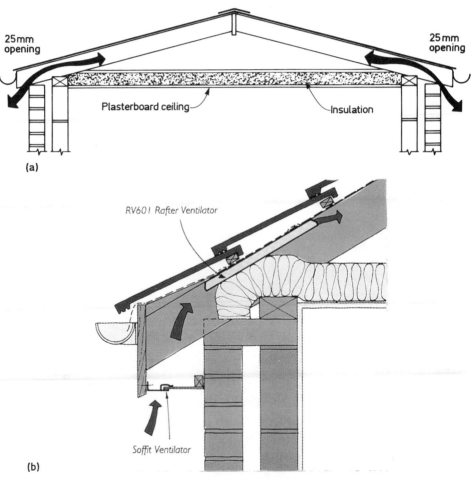

Fig. 19.10 Roof ventilation: (a) duopitch roof below 15°; (b) detail of eaves.

(Part (b) reproduced with permission of Willan Building Services (Glidevale Ltd).)

insulation has to be ventilated. Figure 19.9 provides details of typical ventilator tiles and slates. Figures 19.10 and 19.11 give general ventilation details.

TYPICAL SPECIFICATION FOR TILED ROOF

Refer to Appendix B, clause 14 for details.

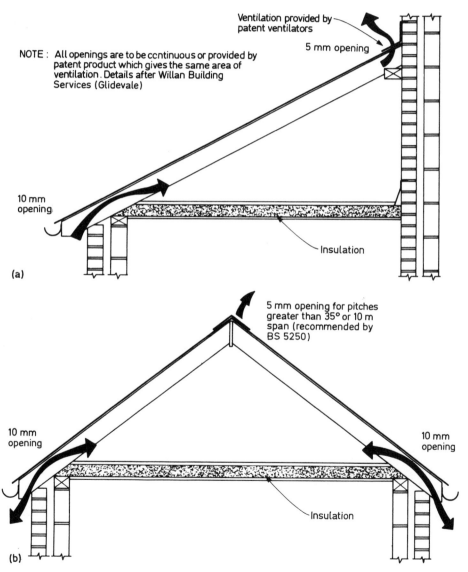

Fig. 19.11 Roof ventilation: (a) monopitch roof over 15°; (b) duopitch roof over 15°.

20 Finishes etc.

GENERALLY

When dealing with small plans, it is unusual to become involved with detailed internal finishes or layouts and I deal with them in one composite description. Refer to standard specification clause 15 in Appendix B.

PLASTERWORK

The following should be borne in mind:

(a) Ceilings with joists up to 450 mm centres – use 9.5 mm plasterboard.
(b) With roof trusses at 600 mm centres – specify 12.5 mm plasterboard.
(c) With the type of external wall construction indicated in Figs. 16.2 to 16.4, lightweight wall plaster should be used to achieve the necessary U-values.

21 Ventilation to habitable rooms, kitchens, bathrooms and WCs

REQUIREMENTS OF THE APPROVED DOCUMENTS

Approved Document F makes very specific requirements with regard to habitable rooms, kitchens, bathrooms and WCs:

(a) A habitable room shall have window openings of 5% of the floor area of the room and a trickle ventilator of 8000 mm^2. Where one room obtains its ventilation via another room (it has no windows of its own) the opening between them must be at least 5% of the floor area of both rooms combined and the room with the window or door in it also has an opening equal in area to 5% of the floor area both rooms combined and trickle ventilation not less than 8000 mm^2 (see Diagram 2 on page 7 of Approved Document F).

(b) Habitable rooms can be ventilated via a conservatory as long as there are ventilation openings in both the conservatory and the habitable room (e.g. patio doors) which have an opening area equal to 5% of the floor area of the conservatory and room added together and trickle ventilation not less than 8000 mm^2 in both the conservatory and the habitable room (see Diagram 3 on page 7 of Approved Document F).

(c) A kitchen has to have an opening window, mechanical extract ventilation and background ventilation. The mechanical extract ventilation has to be capable of extracting moisture laden air at the rate of 60 l/s (but if incorporated within a cooker hood, 30 l/s). The background ventilation should be 4000 mm^2.

(d) A bathroom has to have an opening window, background ventilation of 4000 mm^2 and mechanical extract ventilation capable of extracting air at the rate of not less than 15 l/s which may be operated intermittently.

(e) Sanitary accommodation has to have window openings of 5% of the floor area of the room and background ventilation of 4000 mm^2.

N.B. See Clause 1.5 on page 7 of Approved Document F for non-habitable rooms with no windows.

22 Staircases

REQUIREMENTS OF THE APPROVED DOCUMENTS

Staircases are covered by Approved Document K. I have not included details concerning winding stairs, spiral stairs or tread stairs. (For details on these constructions refer to the document.) On normal staircases Approved Document K dictates the following requirements (Fig. 22.1):

(a) Private stairs (stairs in houses) must not exceed 42° pitch.
(b) Stairs must have a minimum headroom of 2 m. (Slightly less headroom is allowed in loft conversions; see Diagram 3 of Approved Document K for details.)

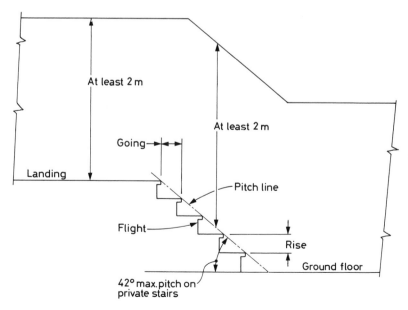

Fig. 22.1 Typical private stairway.

(c) Twice the rise plus once the going should add up to 550–700 mm (maximum rise 220 mm, minimum going 220 mm).
   (i) Any rise between 155 mm and 220 mm used with any going between 245 and 260 mm will comply.
   (ii) Any rise between 165 mm and 200 mm used with any going between 223 mm and 300 mm will comply.
(d) No width given (800–900 mm is normal).
(e) Where a stair has an open side (not between walls) and the drop to the floor is more than 600 mm (or the stair is over 1000 mm wide) a handrail has to be provided. The handrail has to be between 900 mm and 1000 mm high. Upstairs landings should have the same protection.
(f) The balusters to a stair should not allow a sphere of 100 mm diameter to pass through.
(g) Landings must be provided top and bottom of stair flights. The width and length of each landing to be the same as the stair width.

When you are dealing with extensions to existing multistorey properties, it is unusual to become involved in installing a complete new staircase because the house already has one. A very common scenario, however, is the provision of short flights off an existing half landing.

If you refer to Figs. 22.2 to 22.4, I have detailed a typical situation. Fig. 22.2 shows the existing first floor plan and Fig. 22.3 shows the proposed first floor plan which indicates that the house will eventually have two additional

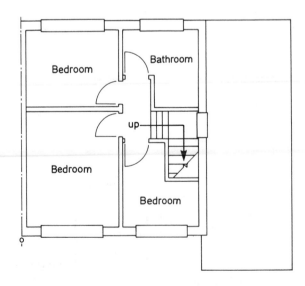

Fig. 22.2 Existing first floor plan.

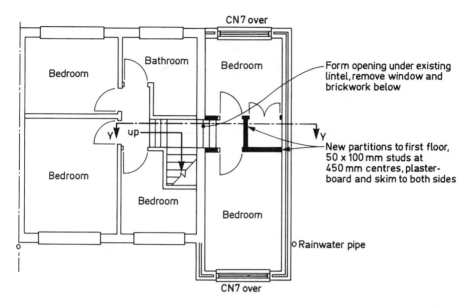

Fig. 22.3 Proposed first floor plan.

bedrooms. The original staircase is not a straight flight but stops at a half landing and then turns and rises again in a short flight until it meets the main landing. As the existing half landing is not at first floor height, there is only one economic solution (unless the existing bathroom or front bedroom is sacrificed) and that is to create a mirror image of the short flight of stairs and create a connection through what was the outer wall. Figure 22.4, section Y–Y indicates the proposal in more detail.

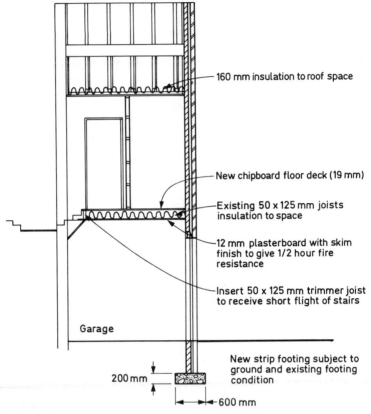

160 mm insulation to roof space

New chipboard floor deck (19 mm)

Existing 50 x 125 mm joists insulation to space

12 mm plasterboard with skim finish to give 1/2 hour fire resistance

Insert 50 x 125 mm trimmer joist to receive short flight of stairs

Garage

New strip footing subject to ground and existing footing condition

200 mm

600 mm

Fig. 22.4 Section Y–Y of proposed first floor.

23 Drainage and plumbing

ABOVE-GROUND DRAINAGE GENERALLY

Figure 23.1 shows a typical rear view of a house with the external pipework on view. This figure shows a typical 'old fashioned' system which has both a hopperhead to take the washbasin discharge and a separate soil and ventilation pipe (SVP) to take the toilet (WC) discharge. Nowadays, hopperheads should not be used and all the waste should be connected to the SVP. As rainwater drainage has already been covered in other sections, the only major item of above ground level drainage is the foul system (i.e. drainage from the WC and the like).

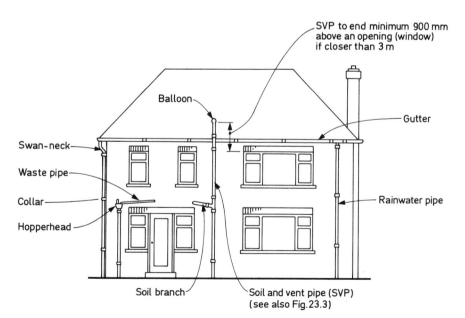

Fig. 23.1 External plumbing (old fashioned system): hopperheads taking upstairs waste should no longer be used.

TRAPS

As you will probably be aware, most sanitary fittings and sinks have a 'trap' underneath (a 'U' bend that holds water). Sanitary appliances and external gullies must have traps provided. The reason why they are there is simple: if they were omitted, foul air from the underground drains would enter the house, which is undesirable. Traps on pipes can be breached, however.

As anyone knows, who has studied physics at school or has ever made wine at home, if you want to get liquid out of one vessel into another it is possible to make a simple syphon with a length of tube. Once the siphonage has started the liquid will empty into the second vessel without any problem as the weight of the water going down the tube (and the effects of atmospheric pressure) forces the liquid out of the first vessel. The principle of siphonage caused a problem for plumbers because sometimes on a long run of pipe, siphonage would start and empty every trap in the house, defeating their purpose (Fig. 23.2). In this sketch a householder has indulged in a little DIY work. Unfortunately, when the water starts to run down the overlong pipe, it is highly likely to syphon off all the water in the trap. In order to overcome this problem there are restrictions on the lengths of pipes (Fig. 23.3).

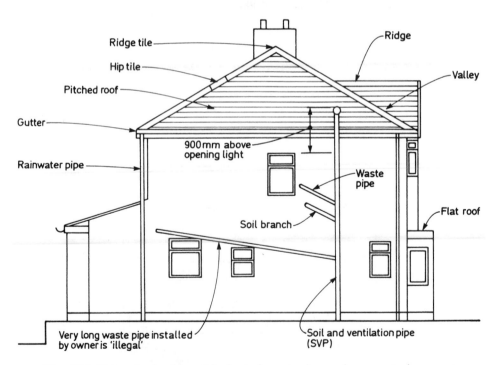

Fig. 23.2 External plumbing (single-stack system): overlong wastepipe causes syphonage problems.

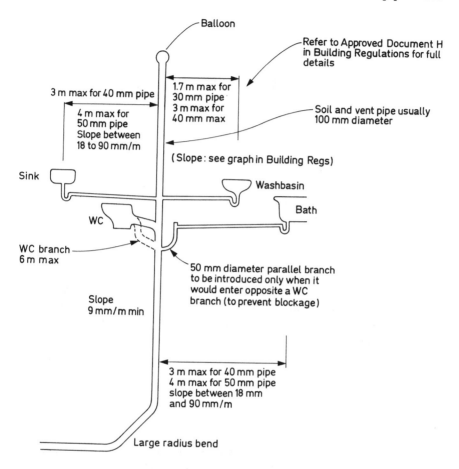

Fig. 23.3 Outline details of single-stack system showing length restrictions on pipes.

WASTE PIPES AND SOIL AND VENTILATING PIPES

It is a requirement of the Approved Document H that below-ground drains are ventilated to the open air so that explosive gases do not build up. A typical soil and vent pipe system is shown in Figs. 23.2 and 23.3.

Most buildings have a combined soil and ventilation pipe because it 'kills two birds with one stone'. The SVP has an open end at high level which will discharge the foul air (the open end should be protected by a 'balloon' grating to stop birds nesting there) and at the same time allows wastes from WCs and wash basins, baths and the like to descend to the below-ground drains. Note that the upper end of all SVP must discharge to the open air in such a way that the foul smells do not come back into the house (i.e. if closer than 3 m to a window, it must extend 900 mm higher than the window.)

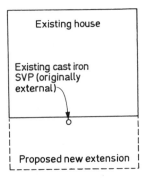

Fig. 23.4 External cast-iron SVP to existing house enclosed by new extension – replace with patent plastic system.

Practical tip

Sometimes a new extension will be built so that the old SVP which was originally outside the building now becomes an internal pipe (Fig. 23.4). Don't just show the pipe boxed in. Indicate that a new plastic patent system will be installed (e.g. Key Terrain, Hunter Plastic etc.) because the old cast-iron type SVP could leak and cause damage at some future date.

SPECIFIC REQUIREMENTS

Above ground

Table 2 of Approved Document H sets out the requirements for traps and pipework which are indicated on Fig. 23.3. If a sketch similar to this one is reproduced on drawings it indicates your intentions to the Building Control Officer and the builder.

Below ground

Refer to clause 15 of Appendix B for standard drainage specification.

24 Other building control aspects

BUILDING A SECOND-STOREY EXTENSION OVER AN EXISTING EXTENSION

Where a new second-storey extension is being built over an older extension, always assuming that the old wall construction is proved to be adequate (e.g. a suitable cavity wall), the Building Control Department will wish to ensure that the existing structure can take the new loads. With this in mind, they will often insist on a note on the plans requiring parts of the old footings to be exposed to prove adequacy.

OLD FOUNDATIONS

Prior to building control being applied to construction work, like most things, foundations were left to the knowledge and skill (or lack of) of the builder/supervising officer. In Figs. 24.1 and 24.2, I have shown typical older

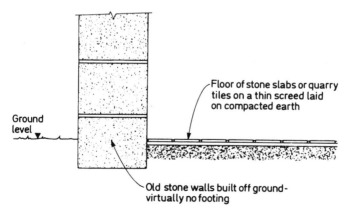

Ground level

Floor of stone slabs or quarry tiles on a thin screed laid on compacted earth

Old stone walls built off ground - virtually no footing

Fig. 24.1 Typical foundation in old cottage.

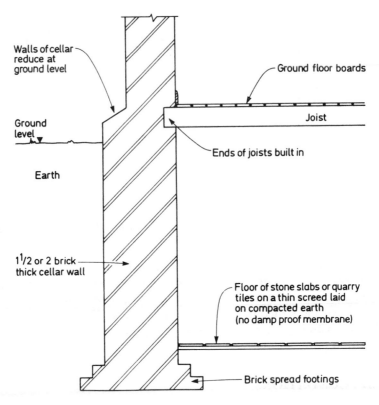

Fig. 24.2 Typical old cellar detail.

types of foundations and the adjacent floor construction. When building near old structures, especially ones that are showing signs of distress, it might be necessary to underpin some foundations (Fig. 24.3). (**Note:** the detail shown on Fig. 24.3 is only an outline sketch. If underpinning work is needed, it may be necessary to consult an engineer.)

THE SLOPING SITE

Sometimes people wish to build an extension or dwelling on a steeply sloping site. When this occurs, there could be problems with the foundations. Approved Document A indicates that the height between the floor slab on one side and the lower ground on the other should not exceed four times the wall thickness. The obvious solution to the problem is to thicken the inside skin of the wall to make the foundations comply or ask an engineer for calculations.

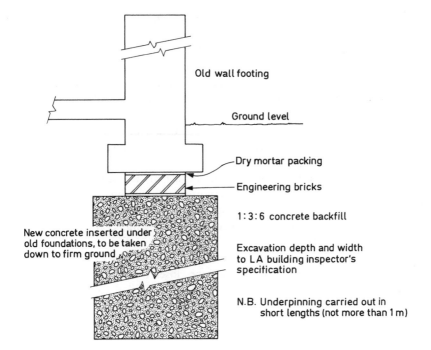

Fig. 24.3 Outline of underpinning detail.

FIRE SAFETY

Approved Document B – 'Fire Safety' is the thickest of the Approved Documents and covers a wide variety of buildings.

Only Section 1 – 'Dwellinghouses' is applicable to this book and covers general provisions, houses with floors more than 4.50 m above ground level and loft conversions.

I have deliberately ignored parts of the Fire Safety document which are unlikely to affect the sort of minor works envisaged in this book.

Note the following:

(a) The Building Control Department cannot force an owner to bring an old building (under former control) up to modern standards unless the proposal is a 'material alteration' or a 'material change of use'. A 'material alteration' is an alteration that if carried out would make the property not comply with modern regulations, or if it did not comply anyway, would make the situation worse.

(b) Where existing houses are 'materially altered', under clause B 0.17, self-contained smoke alarms may have to be installed (see B1 1.8 to 1.15 of Approved Document B). In other words, extending an existing property could result in a smoke alarm system having to be installed. The new alarms have to be mains wired and all alarms have to be linked so that they all go off at the same time. The following requirements are also imposed:

(i) There should be a smoke alarm within 7 m of kitchen/living room doors, measured horizontally (or any other door to a room where a fire is likely to start).

(ii) There should be a smoke alarm within 3 m of all bedroom doors, measured horizontally (or any other door to a room where a fire is likely to start).

(iii) Although battery-powered smoke alarms are widely sold in most department stores, a battery-operated smoke alarm would not comply with modern regulations.

(c) Attached garages that have internal doors that connect them to the rest of the dwelling must have:

(i) a 100 mm step up to restrict burning fluids entering the house.

(ii) A half-hour fire door with a door closer fitted.

PART FIVE: Mainly for Consultants/ Conclusion

25 Preparing to meet the client

What equipment should you take with you when carrying out your survey? I have included a list which covers most eventualities:

(a) Stout A4 clipboard.
(b) Adequate supply of paper (note some people prefer using graph paper as this ensures that sketches are made to scale).
(c) Compass for checking north point.
(d) Variety of coloured pens. (This makes it easier to distinguish dimensions from the outline of the building.)
(e) Instant camera. (This saves a great deal of time for both the draughts-person and for the surveyor when 'on site'. When preparing the plan, you are immediately reminded of what the property looks like and can pick up any items that you may have forgotten to note on your sketches such as rainwater pipe or, say, a balanced flue position.)
(f) Two metre folding surveyor's 'staff'. (I find that a double-sided metric is best.)
(g) Manhole keys (heavy and light duty recommended); screwdriver and crowbar (for lifting manhole covers).

26 Meeting the client

GENERALLY

If you are just reading this book for general interest then this section will probably only receive your passing glance. If you do intend to use the knowledge and take on clients, it is essential that you realize the true implications of your actions. Once you take on your first client, a contract comes into being even on the smallest projects. Your client will treat you as a professional and will expect you to have the necessary expertise to do the job. It is a requirement of most of the well-known professional institutions that surveyors carry professional indemnity insurance. This does not come cheap. Not having insurance is both against the 'rules' and rather foolhardy because we are all human and mistakes can happen. If for some reason you were found negligent, then the costs involved could run into many thousands of pounds.

THE INITIAL ENQUIRY

Although advertising is effective, the majority of enquiries are generated by people who were satisfied with the service offered previously and other personal contacts. This basic principle applies to most businesses whether they be large or small. Obviously, the more satisfied customers that you have, the more likely your business is to thrive. To take matters one stage further, if you look after your clients, like as not, they will come back or recommend your services to others. However, there is always the brand new customer. In order to turn that enquiry into work you need to be able to prove to your prospective client that you provide both an efficient and competitive service. What is more, as most enquiries are by telephone, you have to be able to convince your prospective customer, in a matter of a few minutes, that you can provide the service that he/she requires. In order that you can respond sensibly, you must try to anticipate what information that a client requires.

So what is your prospective client going to ask? The three most common questions that they are likely to ask are:

(a) Can you carry out the service needed?
(b) How much do you charge?
(c) When can you meet to discuss their proposals?

HOW MUCH DO YOU CHARGE?

On the assumption that the answer to question (a) is 'Yes', and that you can meet your client you still have to answer question (b). In some cases, answering that question can be left until you meet. However, most clients want a 'budget'. So how much is a professional service of any sort worth?

It is a question of market demand and competition. It is easy enough to find out what others are charging. My only advice is do not undercut them too severely. It is easy to obtain work by undervaluing your own worth. There are factors which should affect your charges though. Before you can give your prospective client a 'budget' cost you need to know the following:

(a) Where do they live? They could live five minutes away or over two hours drive, and time is money.
(b) What do they want? Your client might just want a single-storey extension, two-storey or a garage.

TALKING THE JOB THROUGH

Once you have arrived at your client's house or the site of the proposed works, your client will obviously describe what he or she wants. Make sure that you note down all the important points. Do not rely on your memory. The important points to remember are:

(a) Agree scope of work. A useful form to carry with you that covers most of the important items to discuss/note whilst on site is shown in Fig. 26.1.
(b) Tell the client what you do and what you don't do. See Appendix D for typical terms and conditions of engagement.
(c) Ask the client to sign a confirmation of instructions form at the time of survey. Obviously, with established clients this might cause offence but with a new customer it should not, as long as you are tactful. He or she is ordering work and you need a confirmation. If you obtain a signed order then a dispute over your agreed fee is unlikely. I have reproduced below a standard form that I carry with me when I visit a client. This is printed out on one of my letterheads (Fig. 26.2).
(d) Confirm your instructions. When sending the copy plan through to the client for approval, I normally send it with a letter on the lines of Fig. 26.3. This letter is then accompanied by a confirmation of instructions (Fig. 26.4).

Another useful form is one called 'What Do I Do Now?' (Fig. 26.5) which I send to clients to keep them updated.

STANDARD SURVEY SHEET

1. Name of Client ....

2. Address of Client ....

3. Owner Y/N

4. If tenant:

   address of owner...

5. Does the alteration affect neighbour by having footings projecting into the next door garden? Y/N

   Is a letter needed? Y/N

NB If you build the extension on the boundary the footings will be under the neighbour's garden. This is acceptable but only if the neighbour agrees (preferably in writing).

6. Name and address of neighbour ....

7. Instructions

   A. ᴱˣᴬᴹᴾᴸᴱ CONSTRUCT REAR EXTENSION TO KITCHEN.

   B. ᴱˣᴬᴹᴾᴸᴱ BLOCK UP OLD SIDE DOOR.

   C. ....

   D. ....

   E. ....

   F. ....

   G. ....

8. Where are drains/manholes? ....

9. Has the garden got over 50% of area left after building extension? Y/N

10. Has the house been extended before? Y/N

If yes explain to client that if existing extension was built after 1948, the cubic capacity counts towards 70/50 m^3.

11. Are there any trees close to the proposed extension? Y/N

Fig. 26.1 Standard survey sheet.

'ON SITE' FORM FOR CLIENT'S SIGNATURE

I confirm that the Conditions of Engagement have been explained to me and that I have been handed a copy for my retention. I accept the estimate figure of £ . . . . . for the service required and understand that this figure will not necessarily be the final cost in the matter.

I enclose my deposit of £ . . . . . and in so doing formally request that you commence the services as outlined.

I understand that I will receive a formal confirmation of instructions and a copy of the plans prior to submission to the Local Authority. Should there be any minor amendments that I require to the plan I understand that these will be made at my request at no extra charge as long as these amendments are made prior to submission.

I understand that before the formal submission can be made I must forward the Local Authority fees as requested in the Confirmation of Instructions.

SIGNED ....

NAME...

ADDRESS ....

POST CODE...................................

DATE ....

TELEPHONE.................................

Fig. 26.2 'On-site' form for client's signature.

Dear Mr and Mrs Jones,

PROPOSED WORK AT YOUR HOUSE

I have pleasure in enclosing a copy of your plan as promised, a copy of my confirmation of instruction and a further copy of my Conditions of Engagement.

Should the plan require any amendments I would appreciate it if you would contact me as soon as possible so that valuable submission time is not lost.

It is essential that you check your plan over carefully, in particular the major dimensions, and ensure that everything is as instructed. I have prepared your plans based upon dimensions taken on site to 'Natural Boundaries' (e.g. fence lines).

If you are in any doubt regarding dimensions indicated on our plan then I would be obliged if you could contact me as soon as possible.

As agreed, my fee does not include for on-site supervision of the works.

Would you please forward the Local Authority fees as soon as possible, and I will then be able to submit the application on your behalf.

I will be forwarding my fee account to you for your attention under separate cover.

Yours sincerely,

A. R. Williams

Fig. 26.3 Covering letter for client's approval of plans.

Confirmation of Instructions

Client: Mr B. Jones

Address: Somewhere, London

Service: Prepare plans to client's approval and submit to Local Authority. No on-site supervision allowed. Agreed fee £ . . . . . (excludes planning and building regulation fee charges; any engineering fee charges which are or may be required by the Building Control Officers; on-site supervision of the works; and VAT of £ . . . . .).

Please forward a cheque for £ . . . . . for Building Control and a cheque for £ . . . . . for Planning. (Please make both cheques payable to Somewhere Borough Council and not to us.)

Note: Our fee account will be forwarded to you under separate cover. The Building Control and Planning charges are to be sent to this office prior to submission of the application. Should the Building Control fee be considered too low or too high, we will notify you and the details of credit/extra charge will be supplied.

Fig. 26.4 Confirmation of instructions.

WHAT DO I DO NOW?

1) EXAMINE YOUR PLANS

Please make sure that you are happy with the plans. Whilst we do try to attain the highest standard, we are only human. If there are amendments needed, please let us know immediately and we will put matters right, if we can.

Should the plan require any amendments we would appreciate it if you would contact us as soon as possible so that valuable submission time is not lost.

It is essential that you check your plan over carefully, in particular the major dimensions, and ensure that everything is as instructed.

We have prepared your plans based upon dimensions taken on site to 'natural boundaries' (e.g. fence lines). If you are in any doubt regarding dimensions indicated on our plan then we would be obliged if you contact us as soon as possible.

2) SEND US THE LOCAL AUTHORITY FEES

On the Confirmation of Instructions we have listed the Local Authority fees that we need to send your application to the Council. Please let us have these fees by return.

3) CAN I START WORK?

NO – we will submit your plans to the Council – it will take several weeks before this has been sorted out. We will send you your Planning/Permitted Development *AND* Building Control details once we receive them.

DO NOT START WORK UNTIL YOU HAVE HEARD FROM US

4) WHAT ARE WE DOING?

By law you need to obtain permission of the Local Authority before you build an extension. Normally this involves two departments which are:
1) Planning
2) Building Control
Once we have received your cheque/cheques for the Local Authority we will send your plans to the Council.

It is unadvisable to start work until these approvals have been received.

5) NOTES

(a) THE DRAWING/SUBMISSION FEES DO NOT INVOLVE ON-SITE SUPERVISION.

(b) COPIES OF CONFIRMATION OF INSTRUCTION AND CONDITIONS OF ENGAGEMENT ENCLOSED.

Fig. 26.5 Form – 'What Do I Do Now?'.

27 Conclusion

GENERALLY

I have tried to cover a large subject in a very small book. Hopefully, I will have succeeded in providing all the necessary basic background information. However, regulations change with time so you must ensure that you keep up to date by studying the technical press and noting new problems as they occur.

As no book can cover every eventuality, I have listed below some useful sources of information (a few have already been mentioned elsewhere in the book).

PROFESSIONAL BODIES

The Royal Institution of Chartered Surveyors
Tel. 0171 222 7000
The Chartered Institute of Building
Tel. 01344 23355
The Association of Building Engineers
Tel. 01604 404121
The Institute of Building Control
Tel. 0181 393 6860

APPROVED INSPECTORS

National House-Building Council (NHBC)
Tel. 01494 434477

DOCUMENTS/BOOKS

The Building Regulations. HMSO
NHBC Standards. NHBC
Ventilation Products. Willan/Glidervale. Tel. 0161 973 1234

COMPUTER SOFTWARE

Pocket Engineer. WL Computer Services. Tel. 0151 426 7400

A FINAL WORD FOR WOULD-BE CONSULTANTS

As I indicated in the previous chapter, if you are just reading this book for general interest, or hope to prepare plans for your own/family use, then such matters as insurance will probably not apply. However, if you are intending eventually to prepare plans for anyone other than direct family/close friends, then you must actively consider the legal implications of your actions. Although professional indemnity insurance is expensive, not having adequate insurance is foolhardy.

APPENDICES

Appendix A   Checklist of items not addressed or often missed off plans

(a) Under the 1976 Regulations a minimum floor-to-ceiling height of 2.30 m was required. The new Approved Documents do not make any requirement for a specific floor-to-ceiling height, except over staircases and under beams (2.00 m minimum and slightly less in loft spaces) and I suppose that if you wanted to design an extension with a ridiculously low floor-to-ceiling height the authorities might be hard pressed to stop you, but I always indicate a minimum headroom of 2.30 m where possible as this is a sensible minimum height for rooms in a dwelling.

(b) The ground floor level/oversite concrete level should not be lower than outside ground level unless special waterproofing precautions have been taken.

(c) The damp-proof course (DPC) in the walls should be minimum 150 mm above ground level.

(d) The floor levels in the extension should be shown to be at the same level as the existing. Steps in floors should only be incorporated if absolutely necessary.

(e) Opening vents in windows. Mark the elevations so the Building Inspector knows which casements open (Fig. 1.1 shows details).

(f) Windows should be described as double glazed.

(g) Ventilation to habitable rooms. (Appendix B, clause 6f gives details.)

(h) When building across a driveway with a garage and there is no other access around the house, try to provide a rear door at the back of the garage (otherwise rubbish will have to be taken through the house).

(i) Windows in an existing house which open into a proposed attached garage should be blocked up to prevent a fire hazard.

(j) As 'rule of thumb' drains should fall to a gradient of 1:40 for 100 mm drains and 1:60 for 150 mm drains (however, note new formulas in Approved Document H).

(k)   Where drains and sewers are close to proposed new foundations ensure that foundations are low enough not to impose any load on the drain or sewer. If a deep drain is near to a wall the trench must be backfilled with concrete.

(l)   Generally, returns to windows should be at least one-sixth of the window opening it abuts (where between windows, one-sixth of combined window). No opening should exceed 300 mm. If it does, engineer's calculations will be required.

(m)   Between attached garages and houses provide half-hour fire doors complete with suitable frame having 25 mm rebates, door closer (e.g. two Perko door closers or one Briton closer). Provide 100 mm step to prevent petrol coming into dwelling in the case of fire.

(n)   When specifying windows, doors, baths, kitchen fittings etc., obtain catalogues (keep them in a filing system for frequent use). Particularly useful catalogues are Magnet and Southerns, and Boulton and Paul. Avoid specifying non-standard sizes.

(o)   Cavity walls: your plans should be drawn in such a way that it is obvious that cavity walls are being used, by drawing in the cavity not only on the sections but also on the small scale plans.

(p)   Cavity closing: note that your drawing should show the cavity closings. Appendix B, clause 4 specifies Thermabate cavity closers. If Thermabate or similar is not used then page 19 of Approved Document L must be observed and the window frame must overlap the blockwork by 55 mm where lightweight plaster is used on the blockwork internally. The blockwork used in the cavity closing must not exceed 0.16 W/m^{2}K and vertical and horizontal damp-proof courses must be shown on the plans.

(q)   Garage walls/porch walls: a half-brick (114 mm thick) wall can be used in garages and porches, but piers as specified in Approved Document A must be provided.

(r)   Roof insulation and ventilation: pitched (tiled and slated roofs) of the cold type must be ventilated to the outside air so that moisture is 'sucked off'. Note the airway sizes shown in earlier chapters.

(s)   Soakers and flashings at abutments: these are specially made pieces of lead (or approved substitute) which fill the gap between the roof tiles and an abutting wall and must be shown on the plans.

(t)   Check drainage and locate manhole. It is a common fault to forget to provide drainage details on the plan.

Appendix B A Standard Specification

The following is a typical example of a standard specification.

1) THIS PLAN IS THE COPYRIGHT of **Andrew R. Williams & Co.** AND IS NOT TO BE REPRODUCED OR USED FOR CONSTRUCTION PURPOSES WITHOUT PERMISSION. It is only for use as a Planning and Building Control document.

BUILDING CONTROL REQUIREMENTS

2) **Generally**

These plans shall not be acted upon until they have been approved in accordance with clauses 13 and 11(1)(b) of the Building Regulations. Should the owner or builder commence work without the above approval they do so at their own risk. In addition, all building construction is to comply in all respects with the requirements of the National House-Building Council's publication *NHBC Standards* (Volumes 1 and 2). All goods and materials unless otherwise specified, shall be in accordance with the latest British Standard and Code of Practice current at the date of tendering. All elements of structure shall have minimum 1/2 hour fire resistance.

3) **Conservation of fuel and power – compliance with Part L 1995**

In accordance with DOE clarification in *IBC News* dated October 1994, all designs are based upon the elemental method with a SAP rating of over 60.

4) **Damp-proof courses and Thermabate cavity closers**

Provide adequate DPCs to new brickwork. DPCs to be a minimum of 150 mm above adjoining ground level. Vertical returns and returns at cill level to window and door openings in new cavity brickwork shall be closed using Thermabate 50 as manufactured by RMC Panel Products Ltd and fixed strictly in accordance with their instructions. (Tel. 0924 362081.)

5) Straps

Eaves level wall plates and flat roof joists shall to be strapped at 1.2 m centres. At floors and roofs, straps shall be provided as indicated in Approved Document A, Diagrams 18a and b and Diagrams 19a, b, c and d. The straps used shall be galvanized mild steel 30 mm × 5 mm × (min.) 1 m long.

6) Insulation/ventilation etc.

6a) Tiled and slated roofs (pitched roofs)

Tiled and slated roofs are to have a U-value of 0.25 W/m^{2}K. This will be achieved by complying with Table A2 of Approved Document L and laying minimum of 160 mm Owens Corning Crown Wool (0.04 W/m^{2}K) between and over the ceiling joists. Cold roof structures shall be ventilated above insulation quilts in accordance with Approved Document F, pages 14 and 15. Roofs under 15 degrees pitch (and those where the ceiling follows the roof) are to have a 25 mm minimum air space at eaves, 5 mm minimum ventilation strip at ridges and 50 mm minimum airspace above insulation. Roofs over 15 degrees – 10 mm minimum airspace at eaves. Lean-to roofs are to have a 5 mm minimum ventilation strip at the higher abutment.

6b) Cold deck flat roofs

Cold deck flat roofs are to have a U-value of minimum 0.35 W/m^{2}K. This will be achieved by laying minimum of 150 mm Owens Corning Crown Wool between the roof joists. Cold roofs structures shall be ventilated above insulation quilts in accordance with Approved Document F, pages 14 and 15 and shall have a 25 mm minimum air space at eaves and 50 mm minimum airspace above insulation.

6c) Warm deck flat roofs

Warm deck flat roofs are to have a U-value of minimum 0.35 W/m^{2}K. The roof deck shall comprise 50 mm of Callenders Metreboard or 90 mm of Roofmax Plus (Owens Corning Insulations) or other equal and approved insulation, laid on a vapour control layer, laid on 12 mm WBP plywood. The voids in a warm deck roof must **not** be ventilated.

6d) Suspended timber ground floors and solid floors in contact with the ground

Note: for insulation of floors see later.

6e) External walls/windows/doors/rooflights

External walls shall have a *U*-value of 0.45 W/m²K. **Note**: for insulation in walls see later. Windows, doors and rooflights shall be composed of either wood or UPVC with a 12 mm airspace and the *U*-value should not exceed 3.3 W/m²K.

6f) Ventilation to habitable rooms

Habitable rooms shall have ventilated openings of at least 1/20th of the floor area of the room. In addition they shall be provided with a controllable background ventilation of not less than 8000 mm² sited at least 1.75 m above floor level.

6g) Kitchens/utility rooms/bathrooms

Kitchens/utility rooms/bathrooms shall have an opening window. In addition they shall be provided with a controllable background ventilation of not less than 4000 mm² sited at least 1.75 mm above floor level and mechanical extract ventilation (60 l/s for kitchens: 30 l/s for utility rooms; 15 l/s for bathrooms.)

7) Foundations

Dimensions given on foundations are only indicative for normal soil conditions. Should it be necessary to provide raft foundation or other construction the builder shall contact the Surveyor as soon as possible. Unless the Building Control Officer instructs otherwise, foundations shall be as follows:

The concrete in the strip foundations shall be composed of a minimum consistency of 50 kg cement to not more than 0.1 m³ of fine aggregate and 0.2 m³ of coarse aggregate) or better (sulphate resisting if necessary). Aggregate to comply with BS 882 Foundations to be taken down a minimum depth of 1.00 m from ground level. The concrete foundations to cavity walls shall be 600 mm wide × 200 mm thick minimum size and for 100 mm thick partition walls 500 mm × 200 mm minimum size. Should the Building Control Officer instruct that differing sizes are to be constructed, the contractor is to comply with such instructions. All foundations to be taken down to good bearing strata and in accordance with Local Authority requirements.

8) Solid floor slabs in contact with the ground under garages, unheated porches, unheated conservatories or extensions to dwellings built prior to 1st April 1990 and under 10 m² of floor area where specificially detailed on the drawings

Uninsulated floor construction shall not be used unless specifically detailed on the drawings. Where used, the construction shall be as follows.

100 mm (min.) concrete floor slab of minimum consistency of 50 kg of cement to maximum 0.11 m^3 of fine aggregate and 0.16 m^3 of coarse aggregate (sulphate resisting if necessary) on 1200 gauge Visqueen laid over blinded hardcore. The surface of the floor slabs shall not be laid lower than the adjacent ground level. If existing levels are inconsistent with this the contractor must lower adjacent levels accordingly.

9) Insulated solid floor slabs in contact with the ground

Insulated construction shall be used and shall achieve U-value of 0.45 W/m^{2}K and shall be constructed as follows.

100 mm (min.) concrete floor slab of minimum consistency of 50 kg of cement to maximum 0.11 m^3 of fine aggregate and 0.16 m^3 of coarse aggregate (sulphate resisting if necessary) on 1200 gauge Visqueen laid over 40 mm Crown floor slab (Owens Corning) or other equal and approved insulation board, on 1200 gauge Visqueen laid over blinded hardcore. In order to avoid cold bridges, the blockwork in the walls directly adjacent to the insulation and floor slab shall be constructed from 100 mm Thermalite Shield 2000 (or other equal and approved construction) as BRE 'Thermal Insulation: Avoiding Risks', Fig. 58. The Visqueen should also be turned up at edges of floor slab to link up with the DPC in the walls. The surface of the floor slabs shall not be laid lower than the adjacent ground level. If existing levels are inconsistent with this the contractor must lower adjacent levels accordingly.

10) Finishes to solid ground floor slabs

50 mm cement and sand screed (1:3) (or 25 mm polished flooring grade asphalt).

11) Suspended timber ground floors

Where detailed on the drawings all ground floor suspended timber floors shall have a subfloor as detailed in 8 above. The surface of the subfloor floor slab shall not be laid lower than the adjacent ground level. If existing levels are inconsistent with this contractor must lower adjacent levels accordingly. Subfloor voids shall be ventilated by air bricks of minimum 1500 mm^2 per open area per linear metre run of wall and sited to maintain a good cross-ventilation. All sleeper walls shall be 'honeycomb' type to allow free air flow. The void height shall be 150 mm minimum to underside of insulation and with a free air flow of at least 75 mm. The floor shall have a minimum U-value of 0.45 W/m^{2}K (this will be achieved by incorporating 60 mm Rocksill thermal slab into the floor, laid on polypropylene netting draped over and between floor joists.

12) External walls

The external walls to new habitable areas shall have a U-value of 0.45 W/ m^2K. All bricks and blocks used in the construction shall comply in all respects with *NHBC Standards*. The external walls shall comprise:

Externally, facing bricks (or thermal blocks and render if applicable) to match existing, and

either 50 mm cavity and 17 mm Celotex CW2000 (or other equal and approved insulant) fixed in accordance with NHBC recommendations and an inner leaf of 100 mm Thermalite Shield 2000 (or other equal and approved block having a U-value of 0.14 W/m^2K and a crushing strength of 4 MN/m^2) or

(after obtaining the prior approval of the Client, Surveyor and the Building Control Officer) an approved construction, complying in respects with the Building Regulations, which will provide a U-value of 0.45 and contain no cold bridges around returned openings. Where the cavity walls are insulated using thermal blockwork thicker than 100 mm, the inner skin of the foundation wall shall be of equal thickness.

Cavities shall contain approved wall ties complying with BS 1243:1972 spaced at 900 mm centres horizontally (max.) and 450 mm vertically (max.) and additional ties at openings. Where cutting and toothing is not possible, 'Furfix' stainless steel connections shall be used. The contractor must ensure that ends of beams are adequately seated on suitable padstones and that lightweight blockwork is not overstressed. The heads of all cavity walls shall be closed using Superlux or similar. Returns around door and window openings shall be closed as described in clause 4 above. All returns in facing brickwork shall be in facings.

13) Flat roofs

The flat roof shall be insulated as described in 6b and 6c above. The waterproof membrane shall comprise 12.7 mm white limestone chippings on 3No. layers of bituminous felt. The three-layer felt system shall comprise a bottom layer to BS 747 type 2B or 3G. The second layer shall be type 3B or 5U and the third layer shall be either type 5B or 5E. The roofing contractor shall comply in all respects with *NHBC Standards*.

Falls are to be created using softwood firrings laid to fall not less than 1 in 40. Softwood joists are to be at maximum 450 mm centres. Provide softwood splayed tilting fillets at edges of the roof in accordance with good building practice.

14) Pitched roofs

All roof pitches shown on the drawings are assessed pitches based upon photographic records taken from ground level and are therefore only indicative of likely pitch (angle of slope). The contractor must calculate the correct pitch from dimensions taken on site. Where a new roof section abuts adjacent tiling the new tiles should match the existing tiles in type, colour and pitch. (Should the existing tiling not be set at recommended pitches the Surveyor should be notified immediately.)

Mortar used on the ridge and verge shall match the colour of the tile. The sarking felt shall be reinforced underslating Andersons Twinplex type IF or other equal and approved, fixed to the soft wood rafters with 2no. galvanized nails size 20 mm into each rafter. The felt is to have 150 mm headlaps, and is to lap 50 mm into the gutter.

Where an existing roof is re-covered the contractor must comply with Section 3 of Approved document A.

Where Russell Grampian smooth or Marley Modern smooth tiles are specified they must not be laid lower than 17.5 degrees. Where Marley Wessex are specified they must not be laid lower than 15 degrees.

Where it proves impossible to lay the tiling at the manufacturers' pitches on any roofing system the Marley specification 'Laying Roof Tiles below Recommended Pitches' shall be followed. In brief the construction below the tiles shall be as follows: 38 mm × 25 mm treated battens on a single layer of BS 747 felt, on 38 mm × 19 mm treated counterbattens, on one layer BS 747 3B felt continuously bonded with 115/15 bitumen on one layer glassfibre-based felt, random tacked on and including 12.5 mm external grade plywood.

15) Leadwork/plumbing/electrical work/drains etc.

The contractor shall install all necessary soakers, flashings, aprons and the like at all abutments sufficient to prevent water entering the building, and all plasterwork, skirtings, architraves, window boards etc.

The contractor shall install all electrical and plumbing work and fittings (if applicable) in strict accordance with the best building practice. Exaxt details of electrical and plumbing (where not detailed) shall be agreed with the client prior to commencement. Electrical installations shall comply with the current edition of the IEE Regulations.

Drains shall be Hepseal or Hepsleeve flexible jointed 100 mm clay pipes with 150 mm beds and surround laid in accordance with Hepworth recommendations to falls of 1:40. Encase all drains under extension with 150 mm concrete. Lintels to be provided in substructure walls where drains pass through foundations as per *NHBC standards*.

New inspection points, chambers and manholes shall be constructed in accordance with the schedules in the Approved Documents. New manholes (where provided) shall comprise 150 mm concrete bed with benching to channels, 255 mm class B engineering bricks in cement mortar 1:3, 150 mm reinforced concrete cover slab and mild steel cover.

16) Planning matters/Permitted Development status

The Contractor must liaise with the client prior to commencing work and ensure that he is familiar with the terms of the Planning consent or Permitted Development Status. Where the Planning Department have made it a condition of approval that certain materials are used, the contractor will be expected to comply with same and provide samples of materials at no extra cost to the client.

Where the extension falls within the Permitted Development (PD) status the contractor must ensure that:

(a) In the case of a terraced house the extension does not exceed 50 m^3.
(b) In any other case 70 m^3.
(c) Any part of the extension which is within 2 m of the boundary does not exceed 4 m in height above ground level.
(d) In the case of porches having PD status the following rules will be observed:
    (i)   The ground area of the structure does not exceed 3 m^2.
    (ii) No part of the porch shall exceed 3 m in height.
    (iii) No part of the porch shall be closer than 2 m from the back of the footway to the road outside.

17) Generally

The terms Builder or Contractor shall mean the person responsible for the construction of the works. The Contractor shall ensure that a responsible person is on site during normal working hours to take instructions. The Contractor is advised to visit the site prior to quoting and to make due allowance when preparing his estimate for access, availability of labour, plant and all things necessary for the construction of the works. No claim will be accepted for want of knowledge at a later date.

The Contractor shall give all notices to Local Authority and Public Undertakings and pay all fees and charges. (NB The Building Control fees for on site inspection of the works will be paid by the client unless agreed otherwise.)

The Contractor shall include for all costs arising from compliance with all Statutory Orders, Regulations, Building Regulations, Bye-Laws and any Acts of Parliament.

The Contractor shall protect the premises during the execution of the work against all damage or vandalism and shall provide tarpaulins and all other necessary coverings, and take adequate precautions to keep new and existing work free from damage by inclement weather during the progress and clear away on completion. The premises are to be secure at the end of each day's work, and the Contractor must reinstate at his own expense any damage caused by neglect in protecting the building. The drawing and Specification is to be read as a whole; if any details whatsoever are not clearly shown or specified the Contractor is to ask for instructions, and if any work be wrongly done, it shall, if the Surveyor so directs, be removed and done again at the Contractor's expense. Site copies of the drawing must be available on the site during the progress of the works for inspection by the Building Control Officer/Owner/Surveyor. All dimensions given whether figured or scaled are to be physically checked on site by the Contractor prior to commencement of work and the Contractor will take responsibility for same.

Any anomalies are to be reported to the surveyor prior to the work being put in hand and prior to ordering materials. If in doubt . . . ask. Figured dimensions are to take preference over scaled dimensions but scaled dimensions are not to be ignored.

The Contractor must furnish the Local Authority with notices of commencement of work and stage of completion and must liaise with water, gas, electricity and British Telecom as necessary and comply with their requirements. The Contractor will be required to maintain and protect all gas and water pipes, electricity cables, sewers etc., and other public property or property of the Local Authority or Public Utility Company which may be encountered during the progress of the works and he shall be responsible for and properly make good any damage to the same to the satisfaction of the Authorities concerned.

Deviations from the drawing can be made but only with consent to the client, building inspector and surveyor.

No part of the work shall be sublet to other persons unless the written authority of the owner and surveyor is obtained.

All materials, appliances, fittings, etc., must be obtained from sources approved by the owner and surveyor with reasonable samples of materials to be used in the work, which samples if approved, shall become a standard of quality.

The contractor is to carefully check the boundaries of the site and is not to build on land not owned by the client without obtaining the neighbours' consent.

The Contractor must indemnify and insure the owner for any damage to persons and/or property for the sum of not less than £500,000 and the contractor will be held liable for any damage or nuisance to, or trespass on the adjoining property arising from or by reason of the execution of the work, and he must take all necessary steps to prevent any such trespass or nuisance being committed.

No trial holes have been taken on site and the Contractor must acquaint himself with the ground conditions of both the site and adjoining areas.

Comply with NHBC Practice note No. 3 'Root Damage by Trees – siting of dwellings and special precautions'.

Allow for keeping the works clear of rubbish during the currency of the Contract and remove from time to time all debris as it accumulates and leave clear and tidy on completion to the satisfaction of the owner.

Appendix C   Building construction terminology

Aggregate   Broken brick, gravel and sand which forms a major part of materials such as concrete and mortar.

Air brick   A perforated brick built into a wall to provide ventilation either to a hidden void (e.g. under a timber floor) or a room.

Asbestos   Fibrous mineral. Airborne fibres are a known health hazard.

Ashlar   Tightly fitting, square cut building stones.

Asphalt   Thick bituminous coating applied hot to waterproof flat roofs, basements etc.

Back inlet gulley (BIG)   Ground level inlets which are connected to the drainage system into which waste and storm water discharges. Unlike a standard gulley, a BIG has an additional inlet (or inlets) to ensure that connections to above-ground drainage can be easily made.

Balanced flue   A duct through a wall which takes air to a boiler from outside and expels waste gases. Usually wall-mounted externally behind modern boiler.

Balustrades   Protective handrails and spindles for staircases and balconies.

Barge board   Similar to a fascia board but runs up the gable of a house covering up exposed roof timbers.

Battens   Thin timber strips to which tiles and slates are fixed.

Benching   Sloping concrete at base of drainage manhole.

Binder   Roof timber running over ceiling joists to provide stiffness.

Birdsmouth   Cut in roof timber to join strut at angle to purlin, wall-plate or other structural timber.

Blockwork (building blocks)   Precast concrete blocks approximately 18 in × 9 in (450 mm × 225 mm). Generally cheaper to build than brickwork. Now a high quality factory-produced product, usually made of thermally insulating material and manufactured by a large number of companies (e.g. Celcon, Thermalite). The term breeze block should not be used as these blocks are no longer made.

Bonding   Method of laying bricks to a regular pattern, e.g. English bond, Flemish bond, English Garden wall bond, stretcher bond.

Borrowed light   Window in interior wall transferring light from outer window.

Brick   Bricks come in various types (e.g. facings (expensive but look nice), commons (cheaper than facings, usually used in foundations), engineering (very hard dense bricks which are used occasionally where high loading or water resisting qualities needed)). Distinct from a building block, usually made of burnt clay but can be made from concrete or calcium silicate. In very approximate terms the modern coordinating brick used in the UK is usually 9 in × 4.5 in × 3 in high (225 mm × 112 mm × 75 mm) and there are approximately 60 bricks per square metre of half-brick wall. For more details see chapter 8.

BS   British Standard.

Building Regulations   Statutory Local Authority control over building works.

Casement   Part of a window.

Cast *in situ*   Material cast on site in formwork – usually reinforced concrete.

Cavity insulation   A cavity in a cavity wall can be insulated using expanded polystyrene or glassfibre bats or blown insulation. Used to increase the thermal efficiency of a wall. Only purpose made, water-repelling insulation should be used (e.g. Dritherm by Pilkington).

Cavity tray   A type of damp-proof course which steps across a cavity wall to ensure that water in the cavity drains towards weep holes in the outer skin of the cavity wall.

Cavity wall   A warm, dry wall used in UK since the 1920s as an alternative to the solid wall. Usually comprises an outer skin of facing brickwork with an inner skin of thermal blocks. Between the two skins there is usually 50 mm gap which stops water passing from the external face to the internal face of a wall. This cavity can be nowadays filled with approved cavity insulation.

Cavity-wall tie (or wall tie)   Purpose-made tie which has been galvanized or made from stainless steel to link the two skins of a cavity wall together. There are many patent types on the market but the standard types are 'butterfly' or 'vertical twist' (Figs. 13.3(b) and 13.3(c)).

Ceiling joist   A joist that only carries the weight of the ceiling and not a floor.

Cement   The term usually refers to ordinary Portland cement as used in the making of concrete.

Cesspit   Non-mains drainage using sealed tank emptied periodically by council.

Cheek   Side face of dormer etc.

Codes of Practice   Various non-statutory recommendations for use of materials.

Collar   Roof timber running between opposing rafters to prevent spread.

Condensation   Water deposit on any surface when critical dew point is reached.

Consumer unit   Modern electric switch box containing fuses or circuit breakers.

Conventional flue   Boiler chimney with boiler air taken from room.

Coping   Brick, stone, or tile finish to top of parapet wall.

Corbel   Projecting support on face of a wall.

Cornice   Decorative plaster moulding at junction of wall and ceiling.

Cowl   Shaped chimney pot used to prevent down-draught.

Cut, tooth and bond   A method of joining old brickwork to new by removing some of the old bricks and lacing in the new so that the new and old structure are adequately tied together.

Dado   Lower part of internal wall approximately 1 m high below timber rail (dado rail).

Damp-proof course (DPC)   Usually a strip of patent water-resisting material incorporated into a wall or around a window or door opening to prevent water penetration.

Damp-proof membrane (DPM)   In modern construction usually a building film such as Visqueen sheet laid below floor slabs to prevent water penetration to the upper surface. The term is also applied to tar-based applications which serve the same function.

Door closer   Usually used on fire doors to ensure that the door returns to its fully closed position in case there is a fire.

Dormer   Window projecting out of roof slope.

Dragon tie   A diagonal piece of timber or metal strap which is fitted internally across the corner of a hipped roof to prevent the roof spreading.

Dry rot   A fungus that destroys timber if the conditions are favourable (Latin name *Serpula lacrymans*). This rot, once established, will travel extensively and force its way through brickwork and plaster to infect new timber.

Eaves   Lowest part of a sloping roof or the area under it.

Efflorescence   Salt deposits on walls or roof tiles where dampness evaporates.

Fabric   Reinforcement wire mesh reinforcement usually used in concrete slabs. Usually specified using a BS code (e.g. A142).

Façade   Front elevation of building.

Fascia board   A piece of boarding that supports the gutters or covers up exposed roof timbers.

Fillet   Triangular sealing of joint between surfaces, generally cement mortar or timber (as in tilt fillet).

Flashing   Lead, zinc, copper or patent strip covering junction of roof slope with wall or chimney.

Flaunching   Fillet of mortar surrounding base of chimney pots.

Flight Straight run of stairs.

Gable Triangular upper part of end wall from eaves level to ridge (sometimes also referred to as 'pike' by bricklayers, but this is a regional term).

Gang-nailed trusses Modern prefabricated roof timbers fixed with plates (e.g. fink truss or monopitch) (Fig. 19.8 shows details).

Going Distance between risers in a staircase.

Gulley Ground level inlets which are connected to the drainage system into which waste and storm water discharges.

Gypsum plasterboard See plasterboard.

High alumina cement (HAC) Must not be used in structural work.

Hardcore Broken brick, broken stone, concrete etc.

Header Brick laid with short end exposed.

Herringbone strutting Cross-shaped layout of timbers nailed between joists as stiffening. It is now possible to obtain purpose-made metal struts (e.g. Expamet struts).

Hip Sloping edge joining two pitched roof slopes.

Honeycomb wall Bricks laid with gaps between to allow ventilation.

Hopperhead Funnel to collect water at top of downpipe.

Invert Bottom of manhole or drain.

Joist Timber support to roof ceiling or floor running parallel with ground.

Lintel (or lintol) Beam or patent steel beam or joist used over opening supporting construction above.

Made ground Potentially difficult sites infilled with hardcore or rubbish.

Mineral felt Common modern flat roof covering with fairly short life.

Monopitch Roof with only one slope.

Mullion Vertical member dividing panes in a window.

National House-Building Council (NHBC) (Tel. 0494 434477). See Chapter 9 for brief details.

Nosing Outer top corner of step or sill.

Oriel window Upper floor window which is cantilevered into a projecting bay.

Outrigger Regional term (Merseyside in particular) usually used when describing a projecting rear extension to a terraced property.

Oversailing course Course of brick or stone projecting out from face of wall (continuous corbel).

Parapet Top of wall carried up above roof or balcony level.

Party wall Each owner owns half with rights in respect of the other half.

Piles Foundation of concrete columns sunk into ground.

Pitched roof Sloping (rather than flat) and covered with tiles, slates etc.

Plasterboard Gypsum plaster sandwiched between two sheets of stout paper.

Plinth Widening at base of wall.

Plumb Vertical.

Ponding Pools of water lying on a flat roof. If water is lying in patches deterioration can result.

Purlin Major roof beam supporting rafters which runs sideways across slope supporting the rafters.

Rafter Sloping roof timber supporting battens and tiles or slates.

Raft foundation Shallow flat reinforced concrete slab used as alternative to a strip footing.

Rendering A surface application of cement and sand (or similar mix) to external face of wall to provide waterproofing.

Retaining wall Supports ground behind and may provide support to structures.

Reveal Return face of corner at window and door openings.

Ridge Top edge of pitched roof.

Rise Vertical distance between stair treads.

Riser Vertical front of a step.

Rising damp Moisture passes up walls and through floors by capillary action.

Rolled steel joist (RSJ) Sometimes used as lintel over openings.

Sarking felt Underfelting used beneath battens and tiles on sloping roofs.

Sash Inner frame to window carrying glass, hence sliding sash windows.

Screed Smooth cement or asphalt finish to concrete floors.

Scrim Hessian-type material used to seal joints in plasterboard.

Septic tank Non-mains drainage using bacterial action to break down sewage.

Serpula lacrymans The Latin name for the dry rot fungus.

Sleeper walls Used beneath timber suspended ground floors to support sleeper plates and joists (e.g. honeycomb sleeper wall).

Soakers Lead or zinc angles between tiles or slates and flashings.

Soffit Underside of eaves behind fascia.

Soil stack (or soil and ventilation stack – SVP). Main vertical drainpipe for WC and other waste water.

Soldier arch Bricks laid vertically on end at top of window or door opening.

Spalling Breaking of surface of tiles or bricks, often due to frost action in winter.

Sprocket Angled timber at foot of rafter to lift roof tiles or slates over gutters.

Stretcher Brick laid sideways, i.e. for single-skin or cavity work.

Strings Sloping boards supporting ends of treads to staircase.

Strip footing Strip of concrete buried in the ground to support a brick wall.

Stucco Architectural term to describe smooth cement rendering used as external finish to walls (also known as render or rendering).

Stud partition Timber-framed walls clad in plasterboard.

Subframe Outer part of window fixed directly to sides of opening.

Subsoil   Material below topsoil on which foundations rest.

Sulphates   Chemicals in the ground which can cause weakening of concretes especially below ground.

Sulphate-resisting cement   Used as an alternative in concrete when subsoil conditions likely to destroy ordinary Portland cement.

Thermoplastic tiles   Floor tiles made from thermoplastic resins.

Threshold   Sill to an exterior door opening.

Timber-frame houses   Built with load bearing timber and (usually) brick face.

Torching   Mortar applied to underside of slates or tiles (an old-fashioned, out-of-date construction technique).

Transom   Horizontal window member separating panes (vertical is mullion).

Trap   Bend in waste pipe prevents air from drain rising (usually found under sinks, wash hand basins etc).

Trussed rafter   See gang-nailed trusses.

Underpinning   Excavating and inserting a new foundation under existing foundation.

Upstand   Vertical face of flashing or soaker or concrete/timber projection.

Vapour check   Barrier to prevent warm damp air entering wall or roof void.

Verge   Edge of pitched roof at gable end.

Vertical damp-proof course   Used at change in level and in basement and around to window and door openings or at a cavity closing.

Weep holes   Allow water to drain from cavity walls and from behind retaining walls.

Wet rot   Fungal attack to woodwork, especially exterior softwood joinery. Not as serious as dry rot.

Appendix D General notes and standard conditions of engagement

DEFINITIONS

1) When reading these conditions the terms you/your refer to the client for whom we are acting.
2) The terms we/us refer to Andrew R. Williams and Co., Chartered Surveyors.

GENERALLY

Before work can commence on site, it is usually necessary to obtain planning and building control approval. Sometimes listed building approval is also required. As part of our brief, we will apply for planning and building control approval for you.

N.B. The plans must NOT be acted upon until they have been approved in accordance with clauses 13 and 11 (1)(b) of the Building Regulations and planning permission/Permitted Development approvals have been obtained. Should you (or your builder) commence work without receiving confirmation from us that the approvals have been obtained then you do so at your own risk.

SERVICE

The normal service carried out on your behalf by us is the initial survey of the property, preparation of outline scheme design drawings at scales 1:100 and 1:50 with sufficient details and information for interpretation of the proposed works for submission to the Local Authorities and for issuing to building contractors to obtain tenders or quotations.

ON-SITE SUPERVISION OF THE WORKS

Our normal charge does not include on-site supervision unless agreed otherwise. You are advised to check the builder's work carefully as it proceeds. If in doubt about any section of work, ask your builder or the Local Authority Building Control Officer. Make sure you get what you really want.

TIME IS NOT THE ESSENCE OF THE CONTRACT

Time is not the essence of the contract. Once the documents are lodged with the council we are powerless to speed up the approvals although we do try to contact the individual Local Authority surveyors at regular intervals in order to assist progress.

PROVISION OF SAMPLES PLAN FOR APPROVAL (ERROR DETECTION)

A sample plan will be provided for your approval approximately 14 days after we have visited your house. Whilst every effort is made to ensure the accuracy of plans it is essential that you examine the plan carefully and ensure that the details are to your satisfaction. If you do not check your plan you could cause yourself a great deal of trouble on site if we have misunderstood your instructions or an error escapes detection. In particular you should check that the major dimensions are as you instructed. No claim will be accepted at a later date for alleged defect or negligence on our part and it is a specific condition of contract that at no time will our liability exceed the value of the fee charged. Once we receive your cheque for the Local Authority fees (see later) this will be taken as approval of the plans (unless you contact us and let us know what revisions are required). There is no charge for revising the plans as long as the changes required are reasonable requests (e.g. there would be no charge for altering a window dimension or door location etc.).

BUILDING CONTROL/BUILDING CONTROL FEES

Most house extensions require Building Control approval and we will advise you regarding the appropriate fee charges levied by the Local Authority. You will be responsible for paying for the Building Control fees and no submission will be made by us until these fees have been received by us.

PLANNING PERMISSION/PLANNING FEES

Most house extensions require planning permission and we will advise you regarding the appropriate fee charges levied by the Local Authority. You will be responsible for paying the planning fees and no submission will be made until these fees have been received by us.

If your property is a listed building or in a special area or if there are tree preservation orders in force, it is essential that you let us know during our visit so that we can make the necessary applications on your behalf. The responsibility for letting us know whether or not the property is listed or in a special area is your responsibility and no claims will be accepted at a later date if this information is not forthcoming. If you are in any doubt you must write/ telephone your council and let us know as soon as possible if these additional consents are needed. Applying for listed building approval etc., should this be required, is an extra service and additional fee may be levied by us for this work.

USE OF THE PLANS/COPYRIGHT

1) The plans are purely for the use of you or your builder and are not for issue to third parties without permission and we will not be responsible for any alleged losses incurred should third parties act upon the details provided.
2) In accordance with the provisions of the Copyright Act 1956 or later amendments, copyright in all drawings remain the property of Andrew R. Williams and Company unless otherwise stated.

FOUNDATIONS DETAILS

3) It is beyond our terms and conditions to dig trial holes (inspection pits to check the subsoil conditions). Should you wish to have trial holes excavated, we will arrange for a local contractor to do so but you will have to bear the cost of this work. Without trial holes being excavated, it is impossible for us to know the exact condition of the substrata in your area. Dimensions given on foundations are therefore only indicative of normal soil conditions. If it is necessary to revise the foundations on site once the works commence this is not a defect of the plans and the plans specifically advise the builder to agree exact foundations details with the Building Control Officer. Should it be necessary to provide calculations or amend the drawings in the light of on-site excavation this is an additional service and a charge will be made.

DRAINAGE DETAILS

4) All drainage shown on the plans is provisional and may require on-site agreement with the Building Control Officer.

DIMENSIONS ON THE DRAWING

5) As we are not supervising the works on site and will not be a party to the contract between yourself and your future builder we make it a condition on our plans that your builder accurately checks all dimensions on site before

starting work and before ordering materials. This is specifically done so that all parties are protected against the possibility of an error getting through the checking procedure (see later for details). Whilst every effort is made to ensure that the dimensions on the plan are as accurate as possible, we are only human and errors can occur. In addition minor variations instigated by yourself, the Local Authority Building Control Officer or your builder, once building works have commenced can have 'knock-on' effects elsewhere in the structure. We would stress that it is essential for you to insist that your builder carries out these checks before carrying out any work. Should the builder need to make amendments you should tell the Building Control Officer before carrying them out. We will not accept any claims for negligence or consequential loss which are the result of non-compliance with the checking procedure.

NATURAL BOUNDARIES

We prepare the plans based upon dimensions taken on site. Where there are no obvious boundaries or they are hidden from view by debris, snow or existing buildings, we will agree the dimensions with you. Where fences and walls exist between properties we accept these 'natural boundaries' as being correct. If you have any doubt regarding the ownership of any land we would advise you to speak to your neighbour and obtain a letter giving their consent to the proposals or contact your solicitor or building society for clarification.

FEE CHARGES

An estimate will be provided, if required, for the likely cost of fee charges. The estimate is based upon our normal hourly charges for surveyor's and draughtsman's time. However, should your requirements or Local Authority requirements create excessive demands on time, we reserve the right to revise charges as necessary. An invoice to cover fees will be issued to you immediately the plans are complete and your remittance should be forwarded with the Local Authority fees unless indicated otherwise on the confirmation of instruction. We reserve the right to charge interest on unsettled invoices at a rate of 2% above the current National Westminster base rate.

FEES NOT INCLUSIVE

Our fees are exclusive of all Local Authority charges (i.e. Planning and Building Control charges) and the cost of consultant engineer's charges for preparation of structural calculations which may be required by Building Control.

DISBURSEMENTS

The cost of printing drawings for your personal use will be charged at a current local printer's rate.

VAT

We are VAT registered and statutory VAT will be added to our fees.

Index

Page numbers in **bold** refer to figures.

Air bricks **87**, 124
Approved Documents 67–9
  A (Structure) 91, 108, 109, 127, 130,
    143, 164, 186, 190
  B (Fire Safety) 165
  E (Sound Resistance) 118
  F (Ventilation) 153–4, 186
  H (Drainage and Waste
    Disposal) 133, 161, 162, 190
  K (Stairways) 155–6
  L (Conservation of Fuel and
    Power) 85–6, 104, 109–110, 122,
    186
Approved Inspectors 69
Area of Natural Beauty 24–5
Article 1 (5) land 25

Bathroom ventilation 154, 187
Binders 143
Block plans 4–5
Blockwork walls 117–18
Bonding (brickwork) 53
Books, reference 179
Bricks
  dimensions 52–3
  matching 52
  secondhand 53
Brickwork
  bonding 53
  cutting and toothing 53, 56–7
  dimensions 49
Building construction terminology 86–9,
  195–200
Building Control
  application form 73–9
  queries 79–80
  rejection notice 80–81
  requirements specification 185–93
  scope of 12, 68–9

Building Control Department/
  Officer 11, 12, 68–9
Building notice system 70
Building Regulations 67–70
Buried services 62

Car port 71
Catnic-type beams 110–12
Cavity closing details **105**, 110, 185
Ceiling joists 143
Cement, sulphate resisting 95
Certificate A (planning application) 36
Certificate B (planning application) 36
Checklist, plans 6, 183–4
Client
  initial enquiry 171–2
  meeting preparations 169
Cold deck roof 129, 135–6, 186
Concrete mixes, foundation 94
Conditions of contract, surveyor's 201–5
Conservation Area 25, 39, **40–46**, 47
Conservatory 71
  ventilation 153
Contractor's responsibilities 191–3
Covered way 71
Cutting and toothing, brickwork 53,
  56–7

Damp-proof course (DPC) 107, 185
Door openings 109–15
Dormer windows 24
Downpipes 133
Drainage
  above ground 159–62
  below ground, *see* Sewers/drains
Dry rot 124–5

Electrical work 190
English bond 54, 103

English garden wall bond 55, 103

Fascia boards 57
Fees
    Local Authority 35, 76, 202
    surveyor's 172, 204
Finishes, floor 188
Fire safety 165–6
Firrings 131–2, **131**, **132**
Five metre rule (Permitted
    Development) 26
Flemish bond 55, **55**, 103
Floor, finishes 188
Floors
    ground 121–3, 187–9
    suspended **122**, **123**, 123–5, 188
    timber 124, 127–8
    ventilation under 124
Footings, tied 63–4, **63**, **97**
Foundations 91–2, 187
    concrete mixes 94
    depth 95
    near drains **100**, 101
    old 163–4
    strip 93–6
    tied (raft) 97
    trench fill 96–7
Fuel and power conservation 85–6, 104,
    185
Full plans application 70, 73–81
Furfix (jointing strip) 56, 57

General Development Order
    (GDO) 23–4
Going (staircase) 156
Ground conditions 91–2
Guttering 133

Habitable room ventilation 153, 187
Hangers 143
Honeycomb sleeper walls 124
'Housefloor' system 122, 124
Householder extension planning
    policies 15–21
Householder planning application
    form 32–3

Information sources 179–80
Initial enquiry 171–2
Inspectors, Approved 69, 179
Instructions, confirming 172, 176
Insulation 186–9
    materials 104

roof 148, 186
Insurance, professional indemnity 180

Kitchen ventilation 152, 187

Landings, staircase 156
Lateral support straps 109, 128
Leadwork 190
Lintels 109–15
Listed buildings 24, 39
Local Authority
    fees 35, 76–9, 202
    planning application form 31–6
    scope of control 11
Location plans **4**, 5

Materials, matching 49, 52–7
Minor works 71
Mortar, matching 53

National House Building Council
    (NHBC) 69
    Standards 70
Neighbours 36, 63–4
Notice 1, Town and Country Planning
    Act 36–7

On-site instruction form 172, 174
On-site supervision 202
Overlooking, see Privacy distance

Partial fill cavity system 104, **105–6**
Party walls 119
Permitted Development (PD) 15,
    23–30, 191
    five metre rule 26
    rights under 27–30
Plan scales 5
Planning control, scope of 12
Planning Department/Authority 11, 12,
    23, 67, 69
Plans 3–6
    block 5
    checklist 6, 183–4
    client approval 172, 175
    location 5
Plasterboard 118, 129–30
Plasterwork 151
Plumbing 190
Porches 24, 71
Privacy distance 64
Professional bodies 179
Professional indemnity insurance 180

Proposals, preparation 7–8
Purlins 143

Raft foundation, *see* Foundations, tied
   (raft)
Rafters 143
   hip 143, 144–5
   valley 143, 144–5
Rejection notice, Building Control 80–
   81
Render finishes 57
Restraint straps 109, 128
Ridgeboard 143
Rise (staircase) 156
Rolled steel joists (RSJ) 112
Roof
   cold deck 129, 135–6, 186
   common faults with flat system 133–5
   finishes (flat roof) 133
   flat 129–33, 189
   insulation 148, 186
   pitched 137, 190
   traditional structure 143–5
   trussed rafter 146, 148
   ventilation 148–9, 186
   warm deck 129, 135–6, 186
Roof coverings
   pitched 137–43, 190
   see also Tiles/slates
Roof decking 132–3
Roof joist, flat roof 130–31
Roof strap 133, 186

Sanitary accommodation
   ventilation 154, 187
Second storey extension 163
Sewers/drains 62, 101, 190–91
Site obstructions 58–64
Slates, *see* Tiles/slates
Sloping site conditions 164
Smoke alarms 165–6
Soil and Ventilation Pipe (SVP) 159,
   159, **160**, 161, **161**, 162, **162**
Specification, typical standard 185–93
Staircases 155–7, **155**
Standard Assessment Procedure
   (SAP) 85–6
Standard Specification 6–7, 185–93
Straps 133, 186
Stretcher bond 54, 103
Strip foundations 93–7

Struts 143
Stud partition walls 117–18
Substrata 91–2
Supervision, on-site 202
Survey sheet 172–3
Surveyor's conditions of contract 201–5
Syphonage problems 160

Terraced houses 25
Thermal bridging 110
Thermal transmittance, coefficient of, *see*
   U-values (thermal transmittance)
Tied footings 63–4, 97
Tiles/slates 137–43
   matching 52
Timber floors 124, 127–8
Timber frame construction 104
Total fill cavity system 104
Traditional structure, roof 143–6
Trap, drainage 160
Tree preservation orders 47
Trees 47, 58
Trench fill foundations 96–7
Trussed rafter, roof 146, 148

U-bend, *see* Trap, drainage
U-values (thermal transmittance) 86,
   107

Vapour check 129–30
Ventilation 186–7
   requirements 153–4
   roof 148–9, 186
   underfloor 124, 188
Vents, drainage 161–2

Wall plates 146
Walls
   blockwork 117–18
   cavity above DPC 104, 107
   cavity below GL 106
   external 103–15, 187, 189
   honeycomb sleeper 124
   internal 117–19
   party 119
   stud partition 118
Warm deck roof 129, 132, 135–6, 186
'What do I do now' form 172, 177
Window openings 109–15
Windows, lining up heads 57